OpenSees 实用教程

古 泉 黄素蓉 著

科学出版社
北京

内 容 简 介

本书深入浅出地介绍了非线性有限元计算软件 OpenSees 的基本建模与编程方法，针对初学者设计了一系列循序渐进的算例，介绍编译和添加新材料方法，帮助使用者快速掌握 OpenSees 使用与编程方法．

本书适合土木工程及相关领域的研究生和科研工作者阅读．

图书在版编目(CIP)数据

OpenSees 实用教程/古泉，黄素蓉著．—北京：科学出版社，2016.12
 ISBN 978-7-03-051808-8

Ⅰ.①O… Ⅱ.①古… ②黄… Ⅲ.①土木工程-应用软件-教材
Ⅳ.①TU-39

中国版本图书馆 CIP 数据核字(2017) 第 031349 号

责任编辑：李 欣／责任校对：邹慧卿
责任印制：吴兆东／封面设计：陈 敬

科学出版社 出版
北京东黄城根北街 16 号
邮政编码：100717
http://www.sciencep.com
北京虎彩文化传播有限公司印刷
科学出版社发行 各地新华书店经销
*

2016 年 12 月第 一 版 开本：720×1000 1/16
2024 年 4 月第八次印刷 印张：12 1/4
字数：232 000

定价：68.00 元
(如有印装质量问题，我社负责调换)

前　言

OpenSees 全称是 Open System for Earthquake Engineering Simulation, 是土木工程学术界广泛使用的有限元分析软件和地震工程模拟平台. 它是由美国太平洋地震研究中心 (Pacific Earthquake Engineering Research Center, PEER) 从 1997 年开始资助, 以加州大学伯克利分校牵头、近十所美国著名高校为主共同开发而成的.

OpenSees 最突出的优点是源代码公开、由学术界共同开发并共享代码, 易于实现学者间深入的科研合作. 研究者无须基于期刊文章重复编制程序代码, 省去这一复杂繁琐的工作, 极大地提高了科研效率. 并且, OpenSees 多年来持续集成包括美国、中国、日本、加拿大、意大利、英国等国的各高校教授自发集成的最新科研成果, 汇集了目前学术界抗震研究的大量最新成果, 为科研工作者提供了一个非常丰富和重要的资源库.

作为新一代的有限元计算软件, OpenSees 致力于强非线性分析, 具有丰富的非线性单元、材料库和针对强非线性分析开发的算法, 可用于分析非线性岩土和结构体系. OpenSees 是美国地震工程模拟网络 (NEES) 的主要计算平台之一, 国内由国家自然科学基金资助的土木工程重大研究计划集成项目 (2012—2015 年) 也将其确定为主要集成平台之一.

OpenSees 另外一个突出特点是使用面向对象的先进程序构架设计, 基于 C++ 实现. 这样, 多所高校开发人员可同时开发. 并且, 基于此框架易于实现并行计算.

OpenSees 还有许多其他优点, 比如内嵌敏感性和优化分析算法; 具有高性能云计算能力, 如 Open Science Grid、TerraGrid 等; 具有由全世界范围教授和学生组成的 OpenSees 学术社区, 有论坛、定期组织学术讨论、培训和问题交流, 成果在维基百科上公布; 通过 OpenFresco 等技术, 能够实现和其他系统的集成以及混合试验等.

多年来, 作为服务于学术界的科研平台, OpenSees 为学术界作出重大贡献. 科研工作者能够直接利用学术最前沿的成果, 更易于实现学术创新; 目前基于此平台在国内外已经培养出超过几十位教授和大量的研究生; 同时由于 OpenSees 社区的学术先进性, 许多学术界新概念都在 OpenSees 上被率先实现和验证, 比如基于性能的计算方法、PBEE 风险评估框架、Hybrid 混合试验方法等; 基于 OpenSees 的很多科研成果已经被工程界广泛接受.

目前, 国内外许多研究团队都把 OpenSees 作为其主要科研平台之一, 在基于 OpenSees 做最新的科研, 比如基于 OpenSees 模拟重大工程强震中的动力响应和破坏过程；研究超高层结构的破坏、倒塌机理以及判别准则；土与结构相互作用；岩土液化分析等.

在这个背景下，本书主要介绍 OpenSees 的基本使用和编程方法，为广大读者提供一个快速入门的方法. 本书的读者对象为土木工程和相关专业的研究生和初学者. 本书分两部分，即 OpenSees 的使用和 OpenSees 编程基础. 第一部分 OpenSees 使用部分包括 OpenSees 下载与运行、算例设计方法、Tcl 基本语法和基本算例等. 算例部分包括梁柱框架结构、土-结构相互作用体系、流固耦合体系、数值优化算例等，另外，还介绍基于 CS 耦合技术将 OpenSees 集成到其他软件的方法以及用 GID 软件为 OpenSees 进行前后处理等. 第二部分为 OpenSees 编程基础，包括 OpenSees 下载与编译方法、C++ 基本语法、OpenSees 添加新材料的方法. 本书提供的算例力求简单易懂、深入浅出，希望用户用最少的时间快速入门. 同时本书的算例也力求"真实"，虽然极简，但单元和材料参数、边界条件、荷载工况等都尽量符合工程实际，希望用户可通过类似模型来计算复杂的实际工程问题.

本书基于作者在厦门大学所教课程的讲义形成. 部分算例在厦门大学建筑学院网站可查到：http://archt.xmu.edu.cn→ 学术研究 →OpenSees 研发. 书中许多内容都得到了 OpenSees 用户的反馈和长期支持，在此表示衷心感谢！本书得到本课题组许多研究生的帮助，包括邱志坚、彭伊、刘永斗、曾志弘、刘轲奇、林纯、李维泉、卢佳盛、张宁、顾晓、施霄勇等，在此一并感谢！另外，本书得到国家科学技术部重点研发项目（编号：2016YFC0701106）和自然科学基金项目（编号：51261120376, 91315301-12, 5157847）资助，特此致谢！

由于本书写作时间较短，虽然经过反复校核，仍然可能存在不足之处，恳请读者谅解，也希望读者能够提供反馈以帮助作者进一步改进.

作 者

2016 年 11 月 20 日于厦门大学

目 录

第一部分 OpenSees 的使用 ··· 1
- 1.1 下载与运行 ··· 1
- 1.2 简单算例设计方法 ·· 4
- 1.3 简单的 Tcl 语法介绍 ·· 9
  - 1.3.1 Tcl 与 OpenSees ·· 9
  - 1.3.2 基本语法 ··· 10
  - 1.3.3 变量 ·· 12
  - 1.3.4 表达式 ··· 13
  - 1.3.5 字符串操作 ··· 15
  - 1.3.6 列表 ·· 16
  - 1.3.7 控制结构 ··· 17
  - 1.3.8 过程 ·· 20
  - 1.3.9 文件操作 ··· 20
- 1.4 框架结构分析 ··· 21
  - 1.4.1 二维弹性柱的静、动力分析 ······························ 21
  - 1.4.2 二维非弹性混凝土门式框架的静力和动力分析 ··············· 28
  - 1.4.3 二维纤维截面混凝土门式框架的静、动力分析 ··············· 36
  - 1.4.4 三维框架结构在地震下的响应分析 ························ 44
- 1.5 土-结构相互作用体系 ·· 55
- 1.6 流固耦合体系 ··· 72
- 1.7 砂土液化模型 ··· 78
- 1.8 数值优化 ··· 84
  - 1.8.1 基于 SNOPT 优化 ······································ 85
  - 1.8.2 实例分析 ··· 87
- 1.9 基于 CS 技术的 OpenSees 耦合计算方法 ······················· 93
- 1.10 OpenSees 的前后处理软件 GID 介绍 ·························· 99
  - 1.10.1 GID 的基本用法 ······································ 99
  - 1.10.2 OpenSees 的问题类型定义 (GID) ······················· 103
  - 1.10.3 OpenSees 的前处理实现方法 ·························· 108
  - 1.10.4 OpenSees 的后处理实现方法 ·························· 109

  1.10.5 实例 ··· 112

第二部分 OpenSees 编程基础 ·· 120

2.1 下载与编译 ··· 120
  2.1.1 下载 OpenSees 源代码 ·· 120
  2.1.2 下载并安装 TCL ··· 122
  2.1.3 下载并安装 Visual Studio 2010 ······································· 123
  2.1.4 测试 Visual Studio 是否安装成功 ····································· 123
  2.1.5 编译 OpenSees 源代码 ·· 125

2.2 C++ 基本语法 ·· 130
  2.2.1 OOP 与 C++ ·· 131
  2.2.2 C++ 基本语法概述 ·· 131
  2.2.3 变量与常量 ·· 133
  2.2.4 表达式与运算符 ·· 135
  2.2.5 函数 ··· 137
  2.2.6 控制程序流程 ··· 139
  2.2.7 数组与指针 ·· 143
  2.2.8 类与对象 ··· 145
  2.2.9 继承 ··· 150
  2.2.10 多态 ··· 153

2.3 OpenSees 添加新材料 ··· 154
  2.3.1 添加新材料背景 ·· 155
  2.3.2 代码修改过程 ··· 156
  2.3.3 建立 Tcl 模型,调试程序 ·· 169

2.4 OpenSees 添加一维理想弹塑性材料 ······································ 172
  2.4.1 添加新材料背景资料介绍 ·· 172
  2.4.2 配置开发环境 ··· 173
  2.4.3 代码修改过程 ··· 173
  2.4.4 建立 Tcl 模型,调试程序 ·· 184

索引 ··· 188

第一部分 OpenSees 的使用

1.1 下载与运行

第一步：进入 OpenSees 官网主页 http://opensees.berkeley.edu/(图 1.1.1)，单击左边菜单栏的 DOWNLOAD 进入下载页面. 需要先注册，该软件是免费的. 单击 registration, 如图 1.1.2 所示. 接着跳转到图 1.1.3 所示的页面，点击 "I agree to these terms".

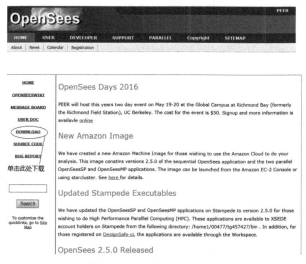

图 1.1.1 OpenSees 主页

图 1.1.2 用邮箱注册 (1)

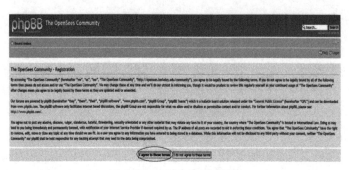

图 1.1.3　用邮箱注册 (2)

第二步：进入图 1.1.4 所示的页面，填写基本信息后点击"Submit"。然后登陆注册所用的邮箱进行激活。

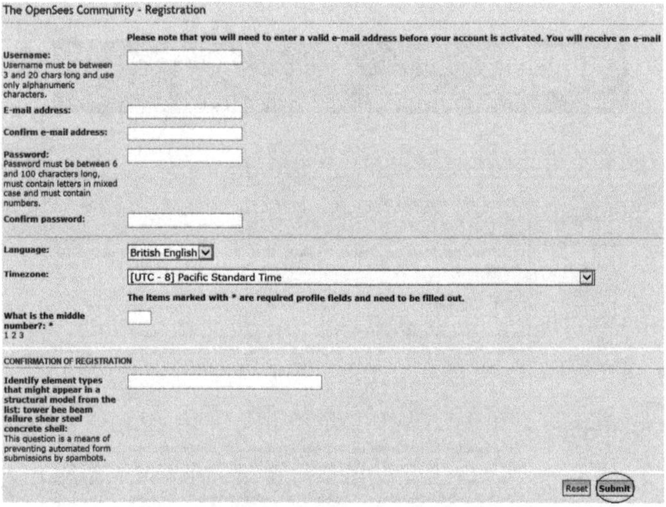

图 1.1.4　用邮箱注册 (3)

第三步：返回 OpenSees 主页进入图 1.1.2 所示的页面，输入已注册的邮箱，点击"Submit"。填写相关信息，如图 1.1.5 所示，点击"Submit"。

第四步：根据所有计算机操作系统的位数，选择安装 32 位或 64 位，先下载安装对应的 tcl/tk 文件，再下载 OpenSees 文件，如图 1.1.6 所示。注意：新版本 (8.5 以后) 的 tcl/tk 文件缺省安装路径目前 "C:\Program Files\Tcl"，旧版本的安装路径为 "C:\Tcl"。建议用户不要修改缺省路径，否则在本书第二部分中编译 OpenSees 源代码时可能出错。

解压下载的 OpenSees 文件，打开 OpenSees.exe 文件，运行界面如图 1.1.7 所示。

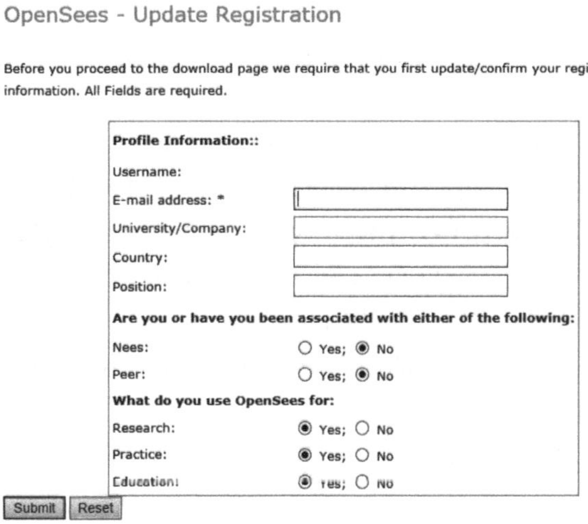

1.1 下载与运行

图 1.1.5 用邮箱注册 (4)

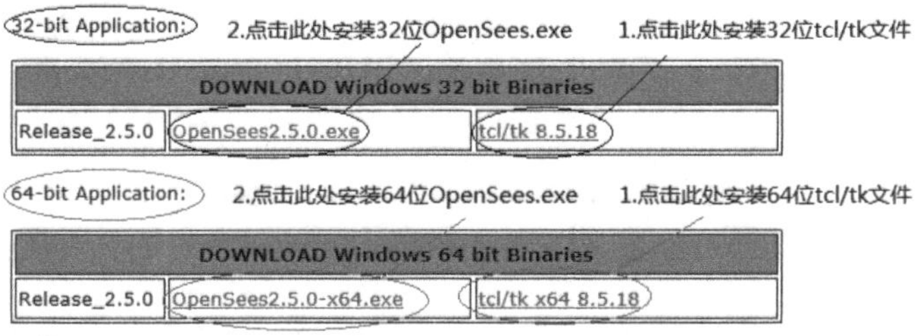

图 1.1.6 OpenSees 和 tcl/tk 的安装顺序

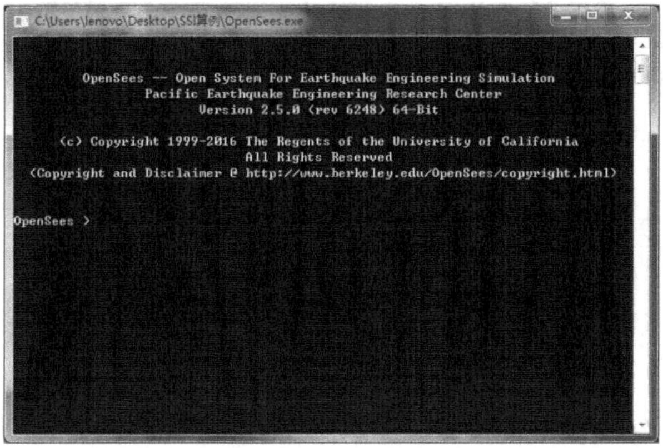

图 1.1.7 OpenSees 2.5.0 运行界面

1.2 简单算例设计方法

下面以一个简单的算例说明使用 OpenSees 建模和分析计算的基本流程. OpenSees 采用 Tcl 语言建模. 读者需要将下面的代码输入文本编辑器, 比如 Notepad 或其他脚本语言编辑器, 然后保存为后缀名为 tcl 的文件. 注意需要将代码前面的行号去掉. 运行时最好将 .tcl 文件和 OpenSees.exe 放在同一个文件夹里, 容易找到文件, 否则需要设置全局文件路径才能找到此文件.

为了说明方便, 对图 1.2.1 中的由三根杆组成的简单结构进行静动力分析. 此处附带 Tcl 代码并解释其意义, 与建模分析相关的 Tcl 语法知识会在 1.3 节进行详细介绍. 为方便解释说明, 代码前加行号, 在实际的模型 tcl 文件中必须删除行号.

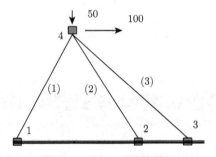

图 1.2.1 杆单元结构算例

OpenSees 有限元分析包括建模和分析两部分. 其中第一部分为建立有限元模型, 即从 1-16 行:

```
1   wipe;
2   model Basic -ndm 2 -ndf 2
3   if { [file exists output] == 0 } {
4       file mkdir output;
5   }
6   node 1 0.0 0.0
7   node 2 144.0 0.0
8   node 3 168.0 0.0
9   node 4 72.0 96.0
10  fix 1 1 1
11  fix 2 1 1
12  fix 3 1 1
13  uniaxialMaterial Elastic 1 3000.0
```

1.2 简单算例设计方法

```
14    element truss 1 1 4 10.0 1
15    element truss 2 2 4 5.0 1
16    element truss 3 3 4 5.0 1
```

以上代码为第一大部分——建模.

第 1 行, wipe 表示删除以往在 OpenSees 里所有模型的信息, 包括节点、单元、材料、边界条件等. 当重复运行建立模型时才用到

> **注意**　wipe 命令不会删除 tcl 中定义的变量 (比如 set a 100, wipe 后 a 变量仍然存在). 利用这点可以做数值优化等分析, 因为设计变量可以保留在 TCL 内存中.

第 2 行 "ndm" 是 "number of dimension" 的缩写, 后面数字 2 表示该模型为二维模型, "ndf" 为 "number of degree of freedom" 的缩写, 后面数字 2 表示模型节点具有 xy 方向两个自由度.

第 3 到 5 行建立输出文件的保存路径, 即结果保存在名为 output 的文件夹里. 这一操作并非必要操作.

第 6 到 9 行建立 4 个节点, 每一行的 node 后面的数字依次为节点编号及该节点的 x、y 坐标.

第 10 到 12 行表示约束条件, 约束 1、2、3 节点在 x、y 方向的位移, 每行后面两个数字 1 表示 x、y 两个方向的位移都约束.

第 13 行表示材料类型为单轴弹性材料, 后面数字 1 表示材料编号 (供后面引用), 3000.0 表示弹性模量为 3000.0. **OpenSees 中单位由用户自己规定, 单位在整个模型中必须统一.**

第 14 到 16 行表示单元类型为桁架单元, 每行第 1 个数字为单元编号, 第 2、3 个数字表示单元的两个端部节点号, 第 4 个数字表示单元截面面积, 最后的数字 1 表示使用标号为 1 的材料. 如第 14 行, 表示桁架单元 1 的两端点为节点 1 和 4, 截面面积为 10.0, 采用 uniaxialMaterial Elastic 材料 (对应标号 1).

OpenSees 模型中还包括输出部分. 和其他有限元不同, OpenSees 并不保存和输出所有节点、单元、材料等信息, 而只输出用户指定的信息. 即从 17-20 行:

```
17    recorder Node -file output/disp_4.out -time -node 4 -dof 1 2
      disp
18    recorder Node -file output/reaction_1.out -time -node 1 -dof 1
      2 reaction
19    recorder Node -file output/reaction_2.out -time -node 2 -dof 1
      2 reaction
20    recorder Node -file output/reaction_3.out -time -node 3 -dof 1
      2 reaction
```

以上代码中关键字 "recorder" 后面的 "Node" 表示记录节点信息 (此外还有 Element 和 Graphics 两大类记录方式), 然后以 "file" 格式保存. 例如第 17 行, 将 4 号节点的位移记录在 output 文件夹下名为 "disp_4.out" 文件里, "time" 表示每计算时步都记录, "-node 4" 表示记录 4 号节点, "-dof 1 2" 表示记录第 1、2 个自由度, 即 x、y 方向, "disp" 表示位移 (displacement). 以此类推, 18-20 行分别记录了 1、2、3 号节点的 x、y 方向反力 "reaction".

OpenSees 中的外力分为结点外力、基底激励 (uniform base excitation) 和多点约束 (multiple support) 三种, 均是通过 pattern 命令来指定. 下面是在节点 4 上施加外力的方法:

```
21    pattern Plain 1 Linear {
22    load 4 100.0 -50.0
23    }
```

第 21 行表示加载模式, 1 为标号. Linear 表示外力是线性增加的外力, 每一步实际外力为系统时间乘以外力系数. 外力系数在第 22 行定义, 即在 4 号节点上加外力系数: $F_x=100$, $F_y=-50$, 系统时间在后面定义. 负号表示与规定的正方向相反.

OpenSees 有限元分析的第二大部分为计算分析, 可以为静力分析或者动力分析 (包括特征分析). 本例中静力分析的 TCL 代码命令流如下:

```
24    constraints Transformation
25    numberer RCM
26    system BandSPD
27    test NormDispIncr 1.0e-6 6 2
28    algorithm Newton
29    integrator LoadControl 0.1
30    analysis Static
31    analyze 10
```

第 24 行表示边界约束方程的处理方式. 除了 Transformation 之外, 还有 Plain 方法、罚函数法等, 详见官网 (http://opensees.berkeley.edu). 本例中三种都可以用.

第 25 行表示结构自由度的编号方式. 除了 RCM 外, 还有 AMD 法、Plain 法等, 详见官网.

第 26 行表示方程的储存和求解方式. 其他方式详见官网.

第 27 行表示用位移增量判断收敛, 后面数字分别表示精度 (最大误差 1.0e−6 此处用绝对误差)、最大迭代数 (6 步)、计算过程在 DOS 屏幕上显示与否所对应的标号 (2 表示只在每时步收敛后输出收敛信息).

第 28 行表示用牛顿迭代法计算. 此外还有修正的牛顿迭代法、牛顿割线迭代法等, 在不收敛时可以替换使用其他算法.

1.2 简单算例设计方法

第 29 到 31 行表示加载方式，LoadControl 表示用力加载控制方式 (此外还有用位移、加速度加载等，详见官网)；0.1 表示每次加载 0.1 倍的外力；具体参考下面 "注意". static 表示静力加载；analyze 10 表示分析 10 步，完成全部外力的加载分析。

计算完毕后，可以检查 4 节点位移，在 OpenSees 提示符下键入 nodeDisp 命令如下：

OpenSees > nodeDisp 4 1
    0.53009277713228364000
OpenSees > nodeDisp 4 2
    -0.17789363846931766000

即可以看到节点 x,y 两个方向位移. 也可以最后查看 output 目录下文件确定位移和支座反力.

注意   本算例中每步 0.1 和 10 步配合使用，完成加载，即每一时步加载 pattern 中定义外力的 0.1 倍，共 10 步完成分析过程. 每一步具体施加外力由 Pattern 决定，本例中每步增加外力为 $F_x = 0.1 \times 100 = 10, F_y = 0.1 \times (-50) = -5$.

更为一般的情况下，外力系数不一定是线性增加，在 Pattern 用 Series 定义外力随时间改变的时程，比如：

pattern Plain 1 {Series -time {0.0 1.0 2.0 3.0} -values {0.0 1.0 0.0 -1.0} } {load 4 100.0 -50.0}

定义了一个随时间改变的外力，具体形式如图 1.2.2 所示. 如果仍然用

integrator LoadControl 0.1
analysis Static
analyze 30

那么 0.1 代表时间步长为 0.1 秒 (虽然是静力分析，OpenSees 中还是有系统时间这一概念)，30 为分析步，即在 3 秒钟内完成此往复加载历程.

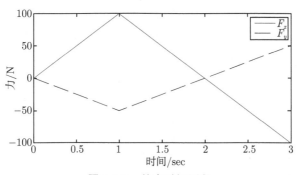

图 1.2.2   外力时间历程

> **注意** 如果在 OpenSees 提示符下键入 "nodeReaction 1 1" 命令来获得支座反力的话，必须先用 "recorder" 命令记录此支座反力 (参考模型第 18 行)，否则系统不会计算此反力，导致输出的反力有误。

OpenSees 动力分析和静力分析命令流很类似. 如果只做动力分析，可以用 32-43 行代替 24-31 行，或者在 31 行后加 "reset" 命令，让系统回到初始状态. 本例中在静力分析基础上做动力分析 (这种工况比较常见，比如先做重力分析，再做地震分析):

```
32   wipeAnalysis
33   loadConst -time 0.0
34   mass 4 100.0 100.0
35   pattern UniformExcitation 2 1 -accel "Series -factor 3 -filePath elcentro.txt -dt 0.01"
36   constraints Transformation
37   numberer RCM
38   system BandSPD
39   test NormDispIncr 1.0e-6 6 4
40   algorithm Newton
41   integrator Newmark 0.5 0.25
42   analysis Transient
43   analyze 2000 0.01
```

以上代码为动力分析部分.

第 32 行表示删除之前的全部分析命令，这里指删除静力分析的 system, numberer, constraints, integrator, algorithm, analysis 等命令.

第 33 行，表示把之前所有静力分析中的外力保持不变，本例中保持以下外力不变: $F_x = 100$, $F_y = -50$. -time 0.0 把此刻系统时间从 1.0 秒重新设为 0.

> **注意** loadConst 命令很重要，在此时刻之后已知外力 (即 pattern Plain 1 Linear) 将不再改变，否则将仍然按照线性增长. -time 0.0 重置系统时间为 0，以免发生错误 (比如后面地震分析中地震动往往从 0.0 秒开始的).

第 34 行表示节点 4 的质量，考虑 x 和 y 方向质量的影响，质量均为 100.0.

第 35 行表示施加 elcentro 地震荷载，用基底激励方式输入. 数字 2 表示荷载标号，因为前面已有重力荷载 (pattern Plain 1 Linear)，因此这里加载标号要区分开，否则会报错; 其后数字 1 表示加速度方向，为 x 方向 (若 2 则表示 y 方向以此类推); 双引号里的内容: 表示读取外部的加速度文件 elcentro.txt，由于 elcentro.txt 文件中没有时间步长，此处指定步长为 0.01 秒，前面的数字 3 表示按加速度文件里的加速度乘以系数 3，作为输入地震动.

1.3 简单的 Tcl 语法介绍

> **注意** 这里的 0.01 秒和后面实际分析所用时步 (43 行) 没有关系. 比如后面 (43 行) 所用步长如果取为 0.005 秒, OpenSees 会自动根据 elcentro.txt 加速度插值得到实际分析步的地震加速度.

第 36~40 行同静力分析.

第 41 行表示用 Newmark 隐式方法计算, 后两个数字分别是 γ、β 值, 此外还可以选择用其他隐式或者中心差分等显式方法计算.

第 42 行表示用瞬态分析方法.

第 43 行表示分析 2000 步, 时间步长为 0.01s.

计算完毕后, 从 disp_4.out 文件读出计算结果 (前面 10 行为静力分析结果), 其中第一列和第二列分别为时间和水平位移, 取动力计算结果 (第 10 行之后)(4 号节点水平位移随时间变化图) 画图如图 1.2.3 所示. (可用 Matlab 的 plot 命令画图)

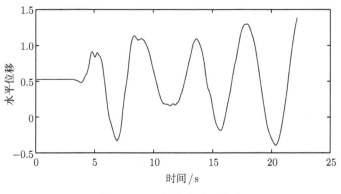

图 1.2.3 4 号节点水平位移

1.3 简单的 Tcl 语法介绍

OpenSees 的前后处理是基于 Tcl 实现的, 本节介绍 Tcl 基本语法. 建议用户把本节的 Tcl 输入计算机快速练习和掌握这一语言 (在 OpenSees 提示符后输入, 即 OpenSees> 之后). **如果暂时不需要编写复杂的 Tcl 计算流程, 用户也可以先跳过这一节, 继续 1.4 节和其后的学习.**

1.3.1 Tcl 与 OpenSees

Tcl(tool command language) 是一种用于控制和扩展应用程序的脚本语言, 由 John Ousterhout 创建. Tcl 有诸多优点, 例如: 语法简单, 容易上手; 跨平台, 兼容多种操作系统; 让应用程序很容易拥有强大的脚本语言支持等. Tcl 语言作为

OpenSees 的前后处理工具, 用来定义分析对象的模型信息、加载方式、方程建立、求解方式、结果记录等.

操作系统作为底层的计算机程序, 是计算机硬件和其他软件的接口, Tcl 语言是基于操作系统的命令解析器, OpenSees 拓展了 Tcl, 增加了有限元的命令解析. 关系如图 1.3.1 所示:

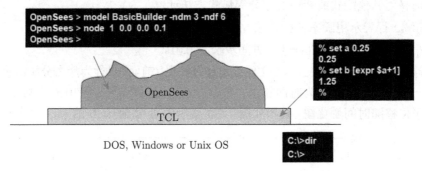

图 1.3.1 操作系统、Tcl 和 OpenSees 的关系

1.3.2 基本语法

◆ Tcl 命令的基本格式

$$\text{command arg1 arg2 ...}$$

第一个单词 command 是命令名, 其他单词 arg1、arg2 为命令的参数, 用空格分隔各个单词.

◆ 基本命令与特殊字符解释

序号	命令	解释
1	set	给变量赋值, 格式为 set var value, 例如, 给变量 A 赋值 2: set A 2.
2	unset	删除一个或多个变量, 释放内存空间.
3	expr	算术运算符
4	puts	输出文本, 多个单词如被空格或 TAB 分隔需要使用 " " 或{} 括起来.
5	info exists	检查变量是否存在, 若存在返回 1, 否则返回 0
6	info global	返回包含所有全局变量名字的一个序列
7	$	变量替代符
8	[]	命令替代符, 括号中的内容为一个指令, 先执行后返回结果.
9	\	反斜杠替代符, 与特殊字母组成转译字符.
10	""	引号可将多个元素组成一个参数, 引号内的内容会被 Tcl 进行置换处理.
11	{}	可将多个元素组成一个参数, 但 Tcl 不对括号中的内容做任何处理.
12	#/;#	注释符号, 在行末命令后为";#", 在行开头"#"和";#"两者均可.

1.3 简单的 Tcl 语法介绍

算例 1.3.1 基本语法和特殊字符使用演示

➤ 在 *OpenSees* 命令提示符后输入(;#后为注释，可不用输入)

```
set E 1                 ;# set 将 1 赋值给变量 E.
set I 1                 ;#set 将 1 赋值给变量 I.
set EI [expr $E *$I]    ;#$进行变量替换,将变量的值替换为 1; [] 执行命令
                         替换, 即把括号内的数字计算出来并替换 [expr $E
                         *$I], 然后赋值给 EI.
puts ''$EI\n''          ;#引号内会进行替换,\n 执行反斜杠替换, 插入换行符.
puts {$EI\n}            ;#花括号内不会进行替换, 原本输出字符.
```

➤ 输出

```
>
1

$EI\n
```

算例 1.3.2 赋值语句 unset 和 info exists 用法演示

➤ 输入

```
set m 2                 ;#内存中定义变量 m=2.
unset m                 ;#删除变量 m.
if {![info exists m]}{  ;#检查是否存在变量m,存在则返回0,不存在则返回 1.
    set m 0
} else {
    set m [expr $m+1]
}
```

➤ puts "$m" 输出

```
>
0
```

算例 1.3.3 info global 用法演示

➤ 输入

```
info global             ;#列出全局变量
```

➤ 输出

>
tcl_rcFileName tcl_version argv argv0 tcl_interactive E auto_path error-Code EI errorInfo auto_execs auto_index I env tcl_patchLevel m argc n tcl_library tcl_plaTform

➢ 输入

puts "$tcl_version"   ;#查询目前 Tcl 版本

➢ 输出

>
8.5

1.3.3 变量

◆ 一般变量

  一个简单 Tcl 的变量包含变量名称和值，名称和值可以是任意字符串，但大小写有区别. 在使用变量前无需事先声明.

◆ 数组

  数组表达一个或多个值到另一个值的映射，它是元素的集合. 每一个元素都有名称和值，元素名称由数组名和数组元素下标名称组成，数组名称和数组元素下标名称可以是任何字符串，包括空格.

◆ 数组操作命令

序号	命令	解释
1	array exists arrayName	判断一个数组是否存在，数组存在返回 1, 数组不存在返回 0
2	array get arrayName	返回数组值的列表
3	array size arrayName	返回数组的大小
4	array set arrayName datalist	定义数组
5	array unset arrayName	删除数组，释放内存空间

算例 1.3.4 set 命令创建数组和数组操作命令演示

➢ 输入

1.3 简单的 Tcl 语法介绍

```
set data_1(name) liming       ;#set 给数组 data_1 下标为 name 的元素设定初值
set data_1(age) 23            ;#set 给数组 data_1 下标为 age 的元素设定初值
set data_1(gender) male       ;#set 给数组 data_1 下标为 gender 的元素设定初值
set data_1(occpuation) work   ;#set 给数组 data_1 下标为 occpuation 的元素
                                  设定初值
set size_data_1 [array size data_1]    ;#返回数组大小
puts "$size_data_1"                    ;#在控制台打印数组 data_1 的大小
puts "[array get data_1]"              ;#取得数组的值列表
array unset data_1                     ;#删除 data_1 数组，释放内存空间
if {[array exists data_1] == 0} {      ;#判断数组 data_1 是否存在
    puts "data_1 is not an array"
} else {
    puts "data_1 is an array"
}
```

➢ 输出

```
>
4
occpuation work gender male age 23 name liming
data_1 is not an array
```

1.3.4 表达式

Tcl 脚本是一组按顺序执行的命令组成，这些命令由表达式和语句构成. 表达式将值和操作符结合起来，运算产生新的值. 操作符包括：算术操作符、关系操作符、逻辑操作符等，这些运算符有不同的优先级别. Tcl 表达式支持常用的数学函数，可以在表达式中直接使用.

♦ 操作符解释

序号	操作符	解释
1	−n	−n:n 的负号
2	n*m n/m	n*m: n 乘以 m, n/m: n 除以 m, n%m: n 除以 m 取余
3	n+m n−m	n+m :n 加 m, n−m :n 减 m
4	~n n<<m n>>m	~n : n 位按位取反, n<<m : n 算术左移 m 位, n>>m : n 算术右移 m 位

续表

序号	操作符	解释
5	n&m n^m n|m	n&m: n 和 m 按位与, n^m: n 和 m 按位异或, n|m: n 和 m 按位或
6	n&&m	n&&m: 逻辑与: 如果 n、m 都为非 0 则返回 1, 否则为 0
7	n||m	n||m: 逻辑或: 如果 n、m 至少有一个非 0 则返回 1, 否则为 0
8	!n	!n : n 为 0 则结果为 1, 否则结果为 0;
9	n>m n<m	n>m: 如果 n 大于 m, 结果为 1, 否则为 0, n<m: 如果 n 小于 m, 结果为 1, 否则为 0
10	n>=m n<=m	n>=m: 如果 n 大于等于 m, 结果为 1, 否则为 0, n<=m: 如果 n 小于等于 m, 结果为 1, 否则为 0
11	n==m n!=m	n==m: 如果 n 等于 m, 结果为 1, 否则为 0, n!=m: 如果 n 不等于 m, 结果为 1, 否则为 0
12	n?m:k	三目运算符, 如果 n 为真, 则执行 m, 否者执行 k

♦ 数学函数

序号	函数	解释
1	abs(n)	取绝对值函数
2	acos(n)	反余弦函数
3	cos(n)	余弦函数 (弧度 n)
4	asin(n)	反正弦函数
5	sin(n)	正弦函数
6	log(n)	自然对数函数
7	log10(n)	以 10 为底的对数函数
8	sqrt(n)	开平方函数
9	exp(n)	指数函数
10	max(a,b,c)	求最大值
11	min(a,b,c)	求最小值
12	tan(n)	正切函数
13	atan(n)	反正切函数

算例 1.3.5　set 命令与输出

▷ 输入

```
set M 64
set N 81
set W [expr "$M + $N"] ;# 变量是否被双引号包含均可, 不过建议使用双引号
set W [expr $M + $N]
set W_LABEL "$M plus $N is "
```

1.3 简单的 Tcl 语法介绍

```
puts "$W_LABEL $W"
puts "The square root of $M is [expr sqrt($M)]\n"
```

➤ 输出

```
>
64 plus 81 is 145
The square root of 64 is 8.0
```

1.3.5 字符串操作

Tcl 的字符串操作功能很丰富,命令包括用于计算字符串的长度、字符串比较、字符串置换等.

♦ 基本字符串操作命令

序号	命令	解释
1	string length	计算字符串的长度
2	string compare	字符串比较,字符串相同,返回 0,第一个字符串在字典中先于第二个,返回 −1,否则返回 1
3	string equal	字符串比较,两个字符串严格相同返回 1,否则返回 0
4	append	字符串粘结函数,将新的项目附加在指定变量内容后
5	format	字符串格式定义函数

算例 1.3.6 字符串操作命令 string length、string compare、string equal 演示

➤ 输入

```
set number_1  2016/                         ;#年份
set number_2  04/16                         ;#日期
set number_3 [append number_1 $number_2]    ;#number_1 变量后加入了 number_2
                                              变量的内容
set Length_1 [string length $number_1]
set Length_2 [string length $number_2]
set Length_3 [string length $number_3]
puts "$number_1"
puts "$number_2"
puts "$number_3"
puts "$Length_1"
```

```
puts "$Length_2"
puts "$Length_3"
if {[string compare $Length_1 $Length_2] == 0} {    ;#对比两者长度是否相同
    puts {Length_1 and Length_2 are equal}
} else {
    puts {Length_1 and Length_2 are not equal} }
if {[string equal $number_1 $number_3]} {           ;#对比两者是否完全相同
    puts {number_1 and number_3 are equal}
} else {
    puts {number_1 and number_3 are not equal} }
```

➤ 输出

```
>
2016/04/16
04/16
2016/04/16
10
5
10
Length_1 and Length_2 are not equal
number_1 and number_3  are equal
```

1.3.6 列表

列表是元素的有序集合，一个列表中可以有任意个数的元素，各个元素也可以任意字符串，比如姓名、年龄、职业等，Tcl 在使用时把列表存放在变量中，传递给命令。

♦ 基本命令解释

序号	命令	解释
1	list arg1 arg2 ...	创建列表，将多个参数构成一个列表
2	lindex list i	操作列表，返回列表第 i+1 个参数的值
3	llength list	返回列表中元素的个数

算例 1.3.7 列表的创建和基本操作演示

➤ 输入

1.3 简单的 Tcl 语法介绍

```
set name_list {liming wangfang chenxiaoli}   ;#set 构造一个名字列表
set age_list {23 42 13 }                     ;#set 构造一个年龄列表
set gender_list {male female female}         ;#set 构造一个性别列表
set data_list [list name_list age_list gender_list]   ;#三者合并,构造一
                                                        个数据列表
set name_list_length [llength $name_list]    ;#[] 返回 name_list 元素个数
set data_list_length [llength $data_list]    ;#[] 返回 data_list 元素个数
puts "$name_list_length"                     ;# 输出 name_list 列表长度
puts "$data_list_length"                     ;# 输出 data_list 列表长度
lindex $data_list 1                          ;# 返回 data_list 第 2 个元素
```

➢ 输出

```
>
3
3
age_list
```

1.3.7 控制结构

♦ 控制结构解释

Tcl 提供了三个用于循环的命令：while、for、foreach，它们的不同之处在于进入迭代前的设置和退出循环的方式.

while 命令需两个参数：condition 和 statement，程序先处理 condition 表达式，如果结果非 0，就执行 statement，循环至 condition 为假时，退出循环.

for 命令需四个参数：initial、test、final 和 statement. 第一个参数 initial 初始化脚本，第二个参数 test 终止循环的表达式判断语句，第三个参数 final 是每执行一次 statement 需要执行的程序，一般为增减计数值. 第四个参数 statement 为构成循环体的脚本.

foreach 命令需三个参数：var、list 和 statement，第一个参数 var 是变量名，第二个 list 是列表，第三个是构成循环体的 Tcl 脚本. foreach 对列表中的每一个元素顺序执行 Tcl 脚本，在每次执行循环块前，foreach 将变量设为列表的下一个元素.

除此之外，Tcl 还提供有条件执行 if 命令，需要两个参数：condition 和 statement，表达式 condition 为真时，执行脚本 statement；否则直接返回. if 命令还可以有多个 elseif 子句和一个 else 子句. 每个子句有相应的表达式和脚本.

♦ 基本格式

序号	命令	基本格式
1	if	if(condition){statement_1
2	if_else	} else if{
3	if_elseif_else	statement_2}else{ statement_3}
4	switch	switch flags value { pattern1 body1 pattern2 body2 ...}
5	while	While(condition){ Statement}
6	for	for {initial} {test }{final}{statement}
7	foreach	foreach Var list {statement}

算例 1.3.8　for 和 foreach 命令使用演示

➢ 输入

```
set sum_1 0
 for {set n 0} {$n<6} {incr n} {          ;#for 求 1 加至 5 的和
   set sum_1 [expr $sum_1 + $n]
}
set sum_2 0                               ;#设置 sum_2 初值
set list [list 1 2 3 4 5]
foreach m $list {                         ;#foreach 遍历列表 list 求和
set sum_2 [expr $sum_2+$m]}
puts "sum_1 is $sum_1"
puts "sum_2 is $sum_2"
if {$sum_1 < $sum_2} {                    ;#注意}{中间留空格
    puts "$sum_2 is greater than $sum_1"
} elseif {$sum_1 > $sum_2} {              ;#注意 elseif 前后留空格
    puts "$sum_1 is greater than $sum_2"
} else {                                  ;#注意 else 前后留空格
   puts "$sum_1 and $sum_2 are equal"
```

}

> 输出

```
>
sum_1 is 15
sum_2 is 15
15 and 15 are equal
```

算例 1.3.9　while 和 switch 命令使用演示

> 输入

```
set n 0;
set sum 0;
while {$n<5} {incr n;set sum [expr $sum + $n] }       ;#while 循环求和
puts "sum is $sum"
switch -- $sum {
    15 {puts "sum is $sum"}
    default {puts "sum is not $sum"}
}
```

> 输出

```
>
sum is 15
sum is 15
```

1.3.8　过程

proc 命令用来定义问题的解决过程, 它使得 Tcl 脚本易于使用. 其基本格式为

$$\text{proc procName argList body}$$

第一个单词 proc 为过程命令, 定义了名为 procName 的过程, argList 是一个过程参数列表, body 是过程块, 包含 Tcl 脚本. 除非用 return 明确指定返回值, 否则返回值一般为最后一行指令执行的结果.

算例 1.3.10　proc 命令使用演示

> 输入

```
proc sum {num_1 num_2 num_3} {                        ;#定义过程 sum
set m [expr $num_1 + $num_2 + $num_3];
```

```
return $m                                          ;#返回和值
}
set a 1
set b 2
set c 3
puts "The sum of $a +$b +$c is [sum $a $b $c]"     ;#调用过程 sum
```

➤ 输出

```
>
The sum of 1 +2 +3 is 6
```

1.3.9 文件操作

Tcl 能够对文件进行操作，比如复制、删除文件、读取文件信息等.

◆ 文件操作命令解释

序号	命令	解释
1	cd	将当前目录转为该参数所指的目录
2	pwd	返回当前工作目录
3	source	读入文件
4	file mkdir	创建新的目录
5	file delete	删除文件
6	file copy	复制文件
7	file exists	如果文件存在返回 1, 否则返回 0

算例 1.3.11 file mkdir、file copy 和 file delete 等命令使用演示

➤ 输入

```
file mkdir Data                         ;#创建 Data 文件夹
logFile Data/data.txt                   ;#文件夹 Data 中创建 data.txt 文件
if { [file exists Data] == 0 } {
    file mkdir Data;
    logFile Data/data.txt
} else {
    file copy Data Data_1                ;#复制 Data 文件到 Data_1
```

```
    file delete Data_1/data.txt      ;#删除 Data_1 文件中的data.txt文件
}
cd Data_1                            ;#进入 Data_1 文件夹目录
pwd                                  ;#显示当前工作目录
```

➤ 输出

```
>
E:/OpenSees 书籍/tcl/Data_1
```

本节介绍了 Tcl 基本语法,用户掌握这些用法,并在下面章节中结合 OpenSees 新加入的命令,比如 node、element 等,可以灵活地建模和分析. 下面章节通过一系列算例,循序渐进、由浅入深地介绍 OpenSees 建模方法. 建议用户输入这些 Tcl,通过独立运行这些算例来学习 OpenSees 建模方法.

1.4 框架结构分析

1.4.1 二维弹性柱的静、动力分析

(一) 问题简述

该算例的模型为弹性混凝土柱,或称竖向悬臂梁,包含 3 个节点、2 个单元,在节点 1 处采用固定端约束,两个单元的弹性模量和面积均相同,截面大小为 $0.5\text{m} \times 0.5\text{m}$,弹性模量为 $3.0 \times 10^{10}\text{Pa}$,模型的几何尺寸如图 1.4.1 所示,模型中使用的单位为:米 (m)、千克 (kg)、秒 (s)、牛顿 (N). 分析模型分别受到如下两种情况的作用时的响应:

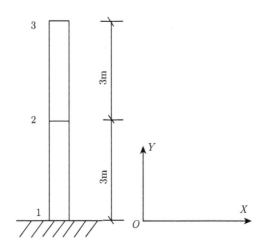

图 1.4.1 弹性混凝土悬臂梁

(1) 重力作用和水平静力 pushover 作用,控制最大水平位移为 0.5m;

(2) 受到最大加速度为 $0.90g(g=9.8\text{m/s}^2)$ 的地震作用.

(二) 命令流分析

(1) 重力作用和水平静力 pushover 作用

```
1   wipe
2   model basic -ndm 2 -ndf 3
3   if { [file exists output] == 0 } {
4     file mkdir output;
5   }
6   node 1 0 0
7   node 2 0 3.0
8   node 3 0 6.0
9   fix 1 1 1 1
10  geomTransf Linear 1
11  element elasticBeamColumn 1 1 2 0.25 3.0e10 5.2e-3 1
12  element elasticBeamColumn 2 2 3 0.25 3.0e10 5.2e-3 1
```

已经在第 2 章讲过的命令将不再赘述.

以上代码为建模部分.

第 2 行表示模型为 2 维,每个节点有 3 个自由度:x、y 和旋转自由度.

第 9 行表示约束条件,约束 1 节点在 x、y 方向的平动位移和平面内的转动位移.

第 10 行表示坐标转换,定义从基本坐标系到整体坐标系的几何变换为线性坐标变换 (即:不考虑几何大变形),变换号编为 1. 基本坐标系、局部坐标系、整体坐标系的具体内容详见 1.4.2 节. 该模型中的整体坐标系和局部坐标系如图 1.4.2 所示.

第 11 到 12 行表示单元类型为弹性梁柱单元,每行第 1 个数字表示单元标号,第 2、3 个数字表示单元的两个端部结点,第 4 个数字表示单元截面面积,第 5 个数字表示单元杨氏模量,第 6 个数字表示单元截面惯性矩,最后的数字 1 表示使用标号为 1 的坐标转换.

1.4 框架结构分析

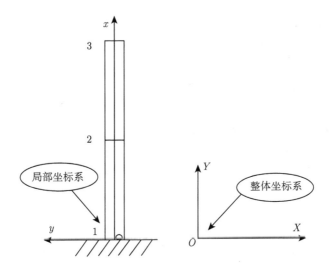

图 1.4.2 弹性混凝土悬臂梁的整体坐标系和局部坐标系

```
13   recorder Node -file output/disp_3.out -time -node 3 -dof 1 2 3
     disp
14   recorder Node -file output/disp_2.out -time -node 2 -dof 1 2 3
     disp
15   recorder Node -file output/reaction_1.out -time -node 1 -dof 1
     2 3 reaction
16   recorder Drift -file output/drift_1.out -time -iNode 1 -jNode
     2 -dof 1 -perpDirn 2
17   recorder Drift -file output/drift_2.out -time -iNode 2 -jNode
     3 -dof 1 -perpDirn 2
18   recorder Element -file output/force_1.out -time -ele 1 global-
     Force
```

以上代码为输出记录部分.

第 16 行 "Drift" 表示输出记录层间位移角. 用第 2 号节点减去第 1 号节点水平位移, 除以 1 与 2 节点间竖向距离以此得到层间位移角.

(1) 重力分析部分

```
19   pattern Plain 1 Linear {
20       load 2 0. -1.0e5 0.0
21       load 3 0. -1.0e5 0.0
22   }
```

```
23    constraints Plain
24    numberer Plain
25    system BandGeneral
26    test NormDispIncr 1.0e-8 6 2
27    algorithm Newton
28    integrator LoadControl 0.1
29    analysis Static
30    analyze 10
31    puts " 重力分析完成..."
```
以上代码为重力分析部分.

> **注意** 对于弹性系统,重力用 1 步加载和分为 10 小步加载是相同的. 但是通常结构体系 (比如高层建筑或者岩土体系) 的弹塑性重力分析需要分多步逐级加载,以避免单步荷载过大造成的不收敛情况.

(2) 水平 pushover 分析

第 20 到 21 行表示在节点 2 和节点 3 上加载,均在 y 方向施加 -1.0×10^5 的力系数. 每一时步具体力的大小为此系数乘以系统时间,本例中每步力的增量为此系数乘以 0.1 秒, 10 个时步内加载完成重力分析. 之后代码为 pushover 分析部分:

```
32    loadConst -time 0.0
33    pattern Plain 2 Linear {
34      load 2 0.5 0.0 0.0
35      load 3 1.0 0.0 0.0
36    }
37    integrator DisplacementControl 3 1 0.001
38    analyze 500
39    puts " 水平力 pushover 分析完成..."
```

第 32 行表示保持重力不变,重新设置时间为 0.

第 33 行到第 35 行表示,在节点 2 和节点 3,均在 x 方向进行加载,加载的比例为 1:2(0.5/1). 由于是位移控制 (第 37 行),此处只能给定力的比例关系,具体力的大小由 OpenSees 计算给出.

第 37 行表示使用位移控制加载,选择节点 3 的 1 自由度的位移进行控制,每时步位移增量为 0.001 米. 由于控制位移, Pattern 2 中力的大小是计算得到的.

第 38 行表示水平推力分析 500 次,则总位移为 0.5m.

1.4 框架结构分析

(3) 受到最大加速度为 $0.90g$ ($g=9.8\mathrm{m/s^2}$) 的地震作用

地震分析通常在重力分析完成以后, 因此只需将 (2) 中 32 行到 39 行代码换成下面的代码即可 (即, 代替水平 pushover 分析部分).

```
40   mass 2 1.0e4 0.0 0.0
41   mass 3 1.0e4 0.0 0.0
42   loadConst -time 0.0
43   timeSeries Path 1 -dt 0.02 -filePath tabas.txt -factor 9.8
44   pattern UniformExcitation 2 1 -accel 1
45   set temp [eigen 1]
46   scan $temp "%e" w1s
47   set w1 [expr sqrt($w1s)]
48   puts " 第一阶频率 f:  [expr $w1/6.28]"
49   set ksi 0.02
50   set a0 0
51   set a1 [expr $ksi*2.0/$w1]
52   rayleigh $a0 0.0 $a1 0.0
53   wipeAnalysis
54   constraints Plain
55   numberer Plain
56   system BandGeneral
57   test NormDispIncr 1.0e-8 10 2
58   algorithm Newton
59   integrator Newmark 0.5 0.25
60   analysis Transient
61   analyze 1000 0.02
62   puts " 地面运动分析完成..."
```

第 40、41 行添加节点集中质量, 动力分析中必须有质量. **OpenSees 中节点质量和单元质量各自计算, 最后累加.** 第 42 行固定重力荷载, 重置时间为 0. 第 43 行定义加载路径, 序列号为 1, 地震加速度文件 tabas.txt 的时间步长为 0.02s, 文件中的加速度矢量值要乘以荷载因子 9.8. 文件 tabas.txt 中的地震的波形如图 1.4.3 所示 (纵坐标为乘以 g 等于实际加速度).

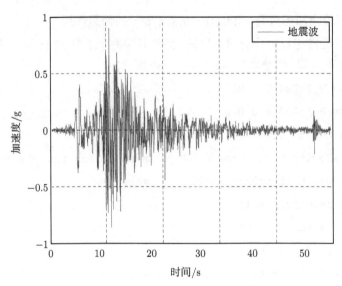

图 1.4.3 tabas.txt 中地震波

第 44 行表示定义荷载形式为基底一致激励, 序列号为 2, 加载方向为第 1 自由度, 即 x 方向, 加载方式为加速度, 定义加速度历史的路径序列号为 1, 即 43 行定义的加载路径.

第 45 行到 52 行根据阻尼比确定 Rayleigh 阻尼的系数 a0 和 a1, 如果用户已经确定了这两个常数, 可以直接赋值并运行第 52 行. 第 45 行的 "eigen 1" 表示求解模型的一个特征值, 返回的是一条包含特征值的 tcl 字符串, 将其赋值给 temp. **特征值为圆频率的平方**.

第 46 行表示将 temp 包含的特征值转换成数字类型, 并将其赋值给 w1s. 第 47、48 行计算频率. 第 49 到 51 行用刚度阻尼公式计算系数 a0 和 a1(参考结构动力学), 第 52 行用 OpenSees 的瑞利阻尼命令定义阻尼.

> **注意** 瑞利阻尼定义中有四个参数: rayleigh \$alphaM \$betaK \$betaKinit \$betaKcomm, 其中第一个参数为质量矩阵的系数 \$alphaM, 后面三个参数分别为本时步当前迭代步的刚度矩阵系数、初始刚度矩阵系数和上一个时步的刚度系数, 最后阻尼矩阵叠加所有这些质量和刚度矩阵的贡献, 即: D = \$alphaM * M + \$betaK * Kcurrent + \$betaKinit * Kinit + \$betaKcomm * KlastCommit, 本例中用初始刚度矩阵.

(三) 结果分析

(1) 重力作用和水平静力 pushover 作用

1.4 框架结构分析

这些计算结果是通过用 Matlab 绘图得到的. 其中, 整体坐标系下悬臂梁的 "pushover 最大分布力-位移" 曲线如图 1.4.4. 图中曲线由输出文件 disp_3.out 中第 1 列记录的 pushover 的最大分布力和第 2 列记录的节点 3 的水平位移得到.

> **注意** 需要说明的是, 当使用位移控制加载时, disp_3.out 中第 1 列记录的不再是时间, 而是力的系数, 即: 此系数乘以 Pattern 2 中的力为实际推力大小.

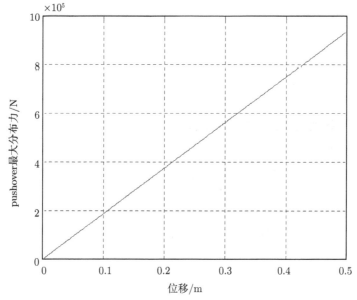

图 1.4.4 pushover 最大分布力-位移响应曲线

(2) 受到最大加速度为 $0.90g(g=9.8\text{m/s}^2)$ 的地震作用

模型水平方向的 "时间-节点水平位移" 曲线: 图 1.4.5 中的 "顶部节点水平位移" 由输出文件 disp_3.out 中第 1 列记录的时间和第 2 列记录的水平方向的位移得到; "中间节点水平位移" 由输出文件 disp_2.out 中第 1 列记录的时间和第 2 列记录的水平方向的位移得到.

> **注意** 第 44 行 "pattern UniformExcitation" 所定义的基底一致输入情况下, 所有 "recorder" 记录的节点响应 (位移、速度、加速度) 都是相对基底的响应 (比如 13-15 行的记录), 而非绝对响应. 当使用 "pattern MultipleSupport" 时, 记录的节点响应都是绝对响应, 在 1.9 节有说明算例.

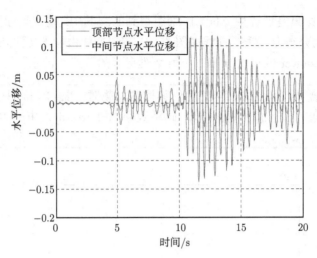

图 1.4.5 节点水平位移-时间

1.4.2 二维非弹性混凝土门式框架的静力和动力分析

(一) 问题简述

该算例的模型为二维非弹性混凝土门式刚架, 柱截面为非弹性组合截面, 截面大小为 $b \times h = 1.52\text{m} \times 1.22\text{m}$, 梁截面为弹性截面, 截面大小为 $b \times h = 1.52\text{m} \times 2.44\text{m}$. 模型的几何尺寸如图 1.4.6 所示, 模型中使用的单位为: 米 (m)、吨 (ton)、秒 (s)、千牛顿 (kN). 求解模型分别受到如下两种情况的作用时的响应分析:

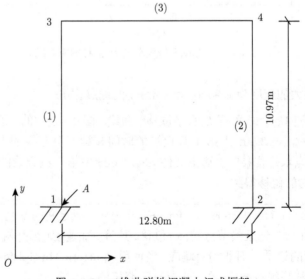

图 1.4.6 二维非弹性混凝土门式框架

1.4 框架结构分析

(1) 重力作用和水平静力 pushover 作用, 控制最大水平位移为 0.5m(不考虑几何大变形);

(2) 受到最大加速度为 $0.90g(g=9.8\mathrm{m/s}^2)$ 的地震作用.

(二) Tcl 命令流分析

(1) 重力作用和水平静力 pushover 作用

```
1   wipe
2   model basic -ndm 2 -ndf 3
3   if { [file exists output] == 0 } {
4       file mkdir output
5   }
6   node 1 0.0 0.0
7   node 2 12.80 0.0
8   node 3 0.0 10.97
9   node 4 12.80 10.97
10  fix 1 1 1 1
11  fix 2 1 1 1
12  uniaxialMaterial Steel01  2  1.47e4  5.74e6  0.01
13  uniaxialMaterial Elastic  3  4.62e7
14  section Aggregator  1  3  P  2  Mz
15  section Elastic  2  2.49e7  3.72  1.8413
16  geomTransf Linear 1
17  geomTransf Linear 2
18  element nonlinearBeamColumn  1  1  3  5  1  1
19  element nonlinearBeamColumn  2  2  4  5  1  1
20  element nonlinearBeamColumn  3  3  4  5  2  2
```

以上代码为建模部分.

第 12 行表示定义单轴材料 Steel01, 算例中用作定义柱的弯曲特性, 第 1 个数字表示材料编号, 第 2 个数字表示屈服弯矩, 第 3 个数字表示弯曲刚度, 第 4 个数字表示刚度比. 材料 Steel01 的 "力-变形" 曲线如图 1.4.7 所示.

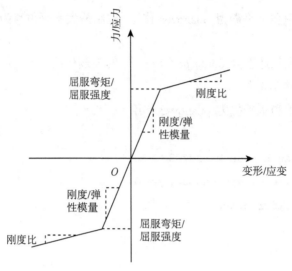

图 1.4.7　Steel01 材料属性

> **注意**　1. 官网中 steel01 各参数的含义：F_y：屈服强度；$E0$：初始弹性模量；b：刚度比.
> 2. 本例中，由于使用了 Aggregator，两个一维材料被用于定义二维截面的轴力-轴向线应变和弯矩-曲率关系. 各模型参数含义变成：对于 2 号材料 (第 12 行)：屈服强度 F_y 变为弯曲强度 M_y、弹性模量 E 变成了抗弯刚度 EI；对于 3 号材料 (第 13 行)：弹性模量 E 变成了抗拉压刚度 EA.

第 13 行表示定义单轴材料 Elastic, 算例中用于定义柱的轴向特性, 第 1 个数字表示材料编号, 第 2 个数字轴向刚度.

第 14 行表示定义非弹性柱截面, 第 1 个数字表示截面编号；第 2 和第 3 个数字表示使用材料的编号, 后面紧跟的字母 "P" 和 "Mz" 分别表示该字母前的材料用于截面的轴向特性和弯曲特性.

第 15 行表示定义弹性梁截面, 第 1 个数字表示截面编号, 第 2 个数字表示弹性模量, 第 3 个数字表示截面面积, 最后一个数字表示截面惯性矩.

第 16 行到 17 行表示坐标转换, 定义从基本坐标系到整体坐标系的几何变换为线性坐标变换 (不考虑几何大变形), 变换号编为 1 和 2. 该模型中的整体坐标系和局部坐标系如图 1.4.8 所示. 由于 2 维框架结构分析不需要指定局部坐标方向, 本算例也可以只定义一个 geomTransf, 即删除 17 行, 三个单元都用同一个 geomTransf.

> **注意**　局部坐标系的 x 方向始终沿杆方向, z 方向垂直纸面向外, 基于此可以确定局部坐标系.

1.4 框架结构分析

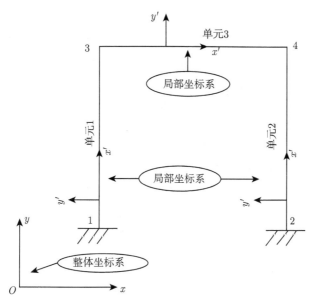

图 1.4.8 二维非弹性混凝土门式框架的整体坐标系和局部坐标系

第 18 到 20 行表示单元类型为非线性力插值梁柱单元, 关于力插值 (而非位移插值) 的理论可参考伯克利的 Filip 教授相关文献. 传统的位移插值欧拉梁用 dispBeam Column 单元, 每行第 1 个数字表示单元标号, 第 2、3 个数字表示单元的两个端部结点, 第 4 个数字表示单元上的高斯点个数, 第 5 个数字表示使用的截面号, 最后的数字表示使用的坐标转换的标号. 下面定义输出文件的记录内容.

21　　recorder Node -file output/disp_34.out-time -node 3 4 -dof 1 2 3 disp

22　　recorder Node -file output/reaction_12.out-time-node 1 2 -dof 1 2 3 reaction

23　　recorder Drift -file output/drift_1.out -time -iNode 1 2 -jNode 3 4 -dof 1 -perpDirn 2

24　　recorder Element -file output/force_12.out -time -ele 1 2 globalForce

25　　recorder Element -file output/foce_3.out -time -ele 3 globalForce

26　　recorder Element -file output/forcecolsec_1.out -time -ele 1 2 section 1 force

27　　recorder Element -file output/defocolsec_1.out -time -ele 1 2 section 1 deformation

```
28   recorder Element -file output/forcecolsec_5.out -time -ele 1 2
     section 5 force
29   recorder Element -file output/defocolsec_5.out -time -ele 1 2
     section 5 deformation
30   recorder Element -file output/forcebeamsec_1.out -time -ele 3
     section 1 force
31   recorder Element -file output/defobeamsec_1.out -time -ele 3
     section 1 deformation
32   recorder Element -file output/forcebeamsec_5.out -time -ele 3
     section 5 force
33   recorder Element -file output/defobeamsec_5.out -time -ele 3
     section 5 deformation
```

以上代码为输出记录部分.

第 24 到 33 行 "Element" 表示输出单元的响应. "-ele 1 2" 表示记录单元 1 和 2 的信息, "section 1" 表示记录的是第 1 个高斯点的响应. "globalForce" 表示以整体坐标系为参考系的力, "force" 也表示以全局坐标系为参考系的力, "deformation" 表示变形. 具体单元记录命令在 1.4.3 有详细讲解.

以下加重力分析.

```
34   pattern Plain 1 Linear {
35       eleLoad -ele 3 -type -beamUniform -122.5
36   }
37   constraints Plain
38   numberer Plain
39   system BandGeneral
40   test NormDispIncr 1.0e-8 6 2
41   algorithm Newton
42   integrator LoadControl 0.1
43   analysis Static
44   analyze 10
45   puts " 重力分析完成..."
```

以上代码为重力分析部分.

第 35 行表示在单元上施加荷载, 在单元 3 上施加沿着局部坐标 y 方向的大小为 -122.5 的均布荷载.

```
46   loadConst -time 0.0
47   pattern Plain 2 Linear {
```

1.4 框架结构分析

```
48   load 3 1.0 0.0 0.0
49   load 4 1.0 0.0 0.0
50   }
51   integrator DisplacementControl 3 1 0.001
52   analyze 500
53   puts " 水平力 pushover 分析完成...!"
```

以上代码为 pushover 分析部分.

(2) 受到最大加速度为 $0.90g(g=9.8\text{m/s}^2)$ 的地震作用

模型受到地震作用时的响应分析与受到 pushover 的区别只出现在重力分析完成以后, 因此只需在节点 3 和节点 4 上加上节点质量并将 (1) 中 45 行以后的代码换成下面的代码即可.

```
54   mass 3 80. 0. 0.
55   mass 4 80. 0. 0.
56   loadConst -time 0.0;
57   timeSeries Path 1 -dt 0.02 -filePath tabas.txt -factor 9.8;
58   pattern UniformExcitation 2 1 -accel 1;
59   set temp [eigen 1]
60   scan $temp "%e" w1s
61   set w1 [expr sqrt($w1s)]
62   puts " 第一阶频率   f: [expr $w1/6.28]"
63   set ksi 0.02
64   set a0 0
65   set a1 [expr $ksi*2.0/$w1]
66   rayleigh $a0 0.0 $a1 0.0
67   wipeAnalysis
68   constraints Plain
69   numberer Plain
70   system BandGeneral
71   test NormDispIncr 1.0e-8 10 2
72   algorithm Newton
73   integrator Newmark 0.5 0.25
74   analysis Transient
75   analyze 1000 0.02
76   puts 地面运动分析完成..."
```

第 67 行 wipeAnalysis 删除静力分析部分 (37-43 行), 其后加动力分析部分 (68-74 行) 进行地震响应分析.

(三) 结果分析

(1) 重力作用和水平静力 pushover 作用

整体坐标系下框架的"pushover 分布力-位移"曲线：图 1.4.9 中曲线由输出文件 disp_34.out 中第 1 列记录的 pushover 的最大分布力和第 2 列记录的节点 3 的水平位移得到.

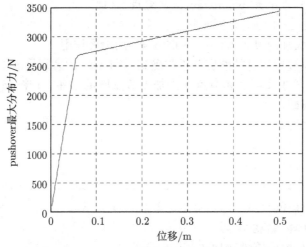

图 1.4.9 pushover 最大分布力-位移响应曲线

局部坐标系下的非弹性混凝土柱 A 点截面的弯矩-曲率曲线：图 1.4.10 中曲线由输出文件 defocolsec_1.out 中第 3 列记录的第 1 个高斯点在局部坐标系下的曲率和输出文件 forcecolsec_1.out 中第 3 列记录的局部坐标系下的弯矩得到.

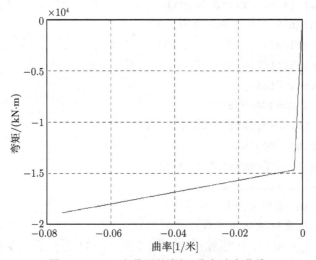

图 1.4.10 A 点截面的弯矩-曲率响应曲线

1.4 框架结构分析

(2) 受到最大加速度为 $0.90g(g=9.8\text{m/s}^2)$ 的地震作用

模型水平方向的"时间-顶点水平位移"曲线：图 1.4.11 中的"顶部节点水平位移"由输出文件 disp_34.out 中第 1 列记录的时间和第 2 列记录 3 节点的水平方向的位移得到.

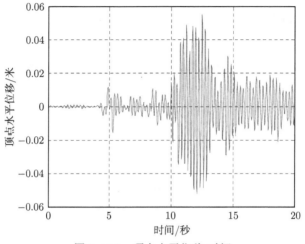

图 1.4.11 顶点水平位移-时间

局部坐标系下的非弹性混凝土柱 A 点截面的弯矩-曲率曲线：图 1.4.12 中曲线由输出文件 defocolsec_1.out 中第 3 列记录的第 1 个高斯点在局部坐标系下的曲率和输出文件 forcecolsec_1.out 中第 3 列记录的局部坐标系下的弯矩得到.

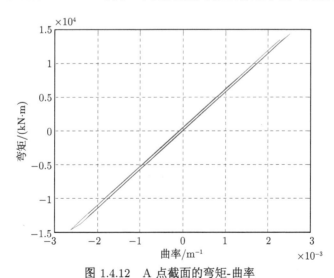

图 1.4.12 A 点截面的弯矩-曲率

> **注意** 本算例中修改第 16-17 行 "geomTransf" 为 "PDelta" 或者 "Corotational"，即可计算几何大变形 (即：小应变大位移). 但是目前 OpenSees 几何大变形还仅限于梁柱框架单元，对于板壳和实体单元都尚未开发. 关于剪力墙构件模型，可参考清华大学陆新征教授网站.

1.4.3 二维纤维截面混凝土门式框架的静、动力分析

(一) 问题简述

该算例的模型为二维纤维截面钢筋混凝土门式框架，柱截面为非弹性纤维截面，截面大小为 $b \times h = 0.5\text{m} \times 0.5\text{m}$，保护层厚度 0.03m，混凝土核心区和保护层采用不同的材料参数，钢筋采用 6 根 25 号钢筋，梁截面为弹性截面，截面大小为 $b \times h = 0.25\text{m} \times 0.6\text{m}$. 模型的几何尺寸如图 1.4.13 所示，模型中使用的单位为：米 (m)、吨 (ton)、秒 (s)、千牛顿 (kN). 求解模型分别受到如下两种情况的作用时的响应分析：

(1) 重力作用和水平静力 pushover 作用，最大水平位移为 0.4m；

(2) 受到最大加速度为 $0.90g(g=9.8\text{m/s}^2)$ 的地震作用.

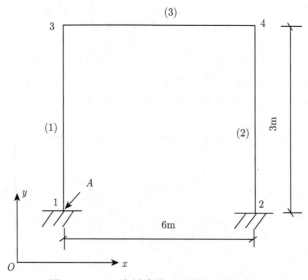

图 1.4.13 二维纤维截面混凝土门式刚架

(二) 命令流分析

(1) 重力作用和水平静力 pushover 作用

```
1  wipe
2  model basic -ndm 2 -ndf 3
3  if { [file exists output] == 0 } {
```

1.4 框架结构分析

```
4    file mkdir output
5  }
6  node 1 0.0 0.0
7  node 2 6.0 0.0
8  node 3 0.0 3.0
9  node 4 6.0 3.0
10 fix 1 1 1 1
11 fix 2 1 1 1
12 uniaxialMaterial Concrete01 1 -34473.8 -0.005 -24131.66 -0.02
13 uniaxialMaterial Concrete01 2 27579.04 -0.002 0.0 -0.006
14 uniaxialMaterial Steel01    3  248200.  2.1e8   0.02
15 section Fiber 1 {
16   patch rect 1  8  8 -0.22 -0.22  0.22  0.22
17   patch rect 2 10  1 -0.25  0.22  0.25  0.25
18   patch rect 2 10  1 -0.25 -0.25  0.25 -0.22
19   patch rect 2  2  1 -0.25 -0.22 -0.22  0.22
20   patch rect 2  2  1  0.22 -0.22  0.25  0.22
21   layer straight 3 3 4.91e-4  0.22  0.22  0.22 -0.22
22   layer straight 3 3 4.91e-4 -0.22  0.22 -0.22 -0.22
23 }
24 section Elastic 2 3.0e7 0.15 4.5e-3
25 geomTransf Linear 1
26 geomTransf Linear 2
27 element dispBeamColumn 1 1 3 5 1 1
28 element dispBeamColumn 2 2 4 5 1 1
29 element dispBeamColumn 3 3 4 5 2 2
```

以上代码为建模部分.

第 12 到第 13 行表示定义单轴混凝土材料 Concrete01, 即改进的 Kent-Scotr-Park 模型, 第 1 个数字表示材料编号, 第 2 个数字表示抗压屈服强度, 第 3 个数字表示峰值应变, 第 4 个数字表示屈服后残余强度, 最后一个数字表示屈服后强度对应的应变. 材料 Concrete01 的本构曲线如图 1.4.14 所示 (图中拉为正). Concrete01 材料不能抗拉.

第 14 行表示定义单轴材料 Steel01, 第 1 个数字表示材料编号, 第 2 个数字表示屈服强度, 第 3 个数字表示弹性模量, 第 4 个数字表示刚度比. 材料 Steel01 的本构曲线如图 1.4.7 所示.

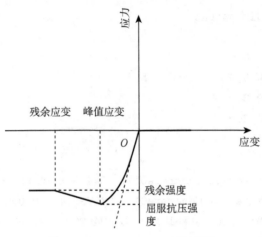

图 1.4.14 Concrete01 材料属性

第 15 行表示定义纤维截面, 标号为 1.

第 16 行到第 20 行的 "patch rect" 表示沿矩形截面定义纤维单元, 参考图 1.4.15 第 1 个数字表示这种纤维使用的材料的标号, 第 2 个数字表示沿局部坐标系的 y 轴划分的子区域个数, 第 3 个数字表示沿局部坐标系的 z 轴划分的子区域个数, 第 4、5 个数字表示局部坐标系下该矩形坐标最小顶点的 y、z 坐标, 第 6、7 个数字表示局部坐标系下该矩形坐标最大顶点的 y、z 坐标. 局部坐标系建立将在下面会详细讲解.

第 21 行到第 22 行的 "layer straight" 表示沿一条直线定义纤维单元 (钢筋), 第 1 个数字表示钢筋使用的材料的标号, 第 2 个数字表示沿这条线上钢筋的个数, 第 3 个数字表示每个钢筋的截面面积, 第 4、5 个数字表示局部坐标系下第一根钢筋的 y、z 坐标, 第 6、7 个数字表示局部坐标系下最后一根钢筋的 y、z 坐标.

第 27 到第 29 行表示单元类型为 displacement beam element, 即位移法插值的欧拉梁单元, 每行第 1 个数字表示单元标号, 第 2、3 个数字表示单元的两个端部结点, 第 4 个数字表示单元上的高斯点个数, 第 5 个数字表示使用的截面号, 最后的数字表示使用的坐标转换的标号.

> **注意** 模型的整体坐标系与局部坐标系的关系, 以及使用 "section Fiber" 定义截面钢筋分布情况可参考图 1.4.15. 局部坐标系的建立在算例 1.4.2 中讲过, 即: 局部坐标系的 x' 方向始终沿杆方向 (比如 "element dispBeamColumn 1 1 3 5 1 1" 指定单元 1 方向从节点 1 到节点 3), z' 方向垂直纸面向外, 基于此可以确定局部坐标系. 图中纤维截面上的 1、2 点表示局部坐标系下的矩形的两个顶点, 据此定义矩形截面大小. 纤维截面始终在 $y'z'$ 面内, 且绕 z' 轴抗弯. 注意不要定义错!

1.4 框架结构分析

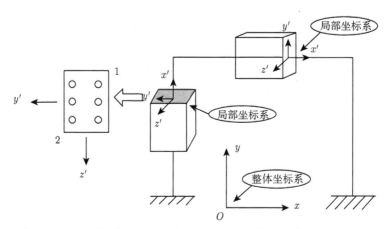

图 1.4.15 二维纤维截面混凝土门式刚架的整体坐标系和局部坐标系

以下为记录输出

30   recorder Node -file output/disp_34.out -time -node 3 4 -dof 1 2 3 disp;

31   recorder Node -file output/reaction_12.out -time -node 1 2 -dof 1 2 3 reaction;

32   recorder Drift -file output/drift_1.out -time -iNode 1 2 -jNode 3 4 -dof 1 -perpDirn 2;

33   recorder Element -file output/force_12.out -time -ele 1 2 globalForce;

34   recorder Element -file output/foce_3.out -time -ele 3 globalForce;

35   recorder Element -file output/forcecolsec_1.out -time -ele 1 2 section 1 force;

36   recorder Element -file output/defocolsec_1.out -time -ele 1 2 section 1 deformation;

37   recorder Element -file output/forcecolsec_5.out -time -ele 1 2 section 5 force;

38   recorder Element -file output/defocolsec_5.out -time -ele 1 2 section 5 deformation;

39   recorder Element -file output/forcebeamsec_1.out -time -ele 3 section 1 force;

40   recorder Element -file output/defobeamsec_1.out -time -ele 3 section 1 deformation;

```
41   recorder Element -file output/forcebeamsec_5.out -time -ele 3
     section 5 force;
42   recorder Element -file output/defobeamsec_5.out -time -ele 3
     section 5 deformation;
```

以上代码为输出记录部分。

> **注意**　用于记录单元的命令 recorder Element 比较复杂，其后的关键词和命令格式因单元而异（即：-ele 3 后面可以用的关键词，比如 "-ele 3 globalForce" 或者 "-ele 3 section 1 deformation"）。目前官网还没有对每一种单元都很好的注释。为了能够正确使用用于记录单元的 recorder 命令，建议用户到此单元的源代码中查看其 setResponse() 函数。比如本例为二维非线性梁柱单元 dispBeamColumn：http://opensees.berkeley.edu/→Developer→Browse the Source Code→trunk/SRC/element/dispBeamColumn/DispBeamColumn2d.cpp，查找到 Response*DispBeamColumn2d ::setResponse(..) 函数，看到 argv[0] 中存的就是这种单元可以识别的关键词，比如 "basicStiffness" "globalForce" "localForce"，等都是合法的关键词。
>
> 我们还可以看到 "globalForce" 所对应的数字为 1，即：theResponse = new ElementResponse(this, 1, P)。具体记录内容从另一个函数 int DispBeamColumn2d::getResponse(..) 中可以查到，数字 "1" 对应输出内容如下：
>
> ```
> if (responseID == 1)
>     return eleInfo.setVector(this->getResistingForce());
> ```
>
> 即输出单元在全局坐标下的内力。
>
> 如果想要进一步记录此单元的某个高斯点的信息（比如本例 "-ele 3 section 1 deformation"），可以继续查看函数 DispBeamColumn2d::setResponse(..) 后面部分："else if (strstr(argv[0],"section") != 0) …" 中的内容，可以看到这里规定其后第一个数字 argv[1] 为高斯点号（高斯点编号沿杆方向从 1 开始编号），本例中 section 后的数字 "1" 表示第 1 个高斯点，OpenSees 将关键词其余部分（本例中为 "deformation" 关键词）转到高斯点上继续解析，高斯点为此单元所用的截面或者材料，本例中为 2D fiber section，即：/trunk/SRC/material/section /FiberSection2d.cpp，其 setResponse(..) 函数规定了合法的关键词（本例中为 "deformation" 关键词）。用这种方法可以非常清楚地知道 recorder 单元的写法和记录的内容。
>
> 还需要指出的是，OpenSees 中梁柱单元除了局部坐标和全局坐标外，还定义了基本坐标系，在此系中单元只有 3 个独立自由度：两个节点的转动自由度和轴向伸缩自由度。基于虚功原理可以得到这三个自由度与局部和全局坐标（均为 6 个自由度）之间的关系。具体可以参考结构分析（structural analysis）相关书籍。

1.4 框架结构分析

下面定义静力加载方式. 由于和前面相同, 不再详细解释.

```
43  pattern Plain 1 Linear {
44     eleLoad -ele 3 -type -beamUniform -65.33;
45  }
46  constraints Plain;
47  numberer Plain;
48  system BandGeneral;
49  test NormDispIncr 1.0e-8 6 2;
50  algorithm NewtonLineSearch 0.75;
51  integrator LoadControl 0.1;
52  analysis Static
53  analyze 10;
54  puts " 重力分析完成..."
```

以上代码为重力分析部分.

```
55  loadConst -time 0.0;
56  pattern Plain 2 Linear {
57     load 3 1.0 0.0 0.0;
58     load 4 1.0 0.0 0.0;
59  }
60  integrator DisplacementControl 3 1 0.001;
61  analyze 300;
62  puts " 水平力分析完成..."
```

以上代码为 pushover 分析部分.

(2) 受到最大加速度为 $0.90g(g=9.8\text{m/s}^2)$ 的地震作用

模型受到地震作用时的响应分析与受到 pushover 的区别只出现在重力分析完成以后, 因此只需在节点 3 和节点 4 上加上节点质量并将 (1) 中 56 行以后的代码换成下面的代码即可.

```
63  mass 3 20. 0. 0.
64  mass 4 20. 0. 0.
65  loadConst -time 0.0;
66  timeSeries Path 1 -dt 0.02 -filePath tabas.txt -factor 9.8;
67  pattern UniformExcitation 2 1 -accel 1;
68  set temp [eigen 1]
69  scan $temp "%e" w1s
70  set w1 [expr sqrt($w1s)]
```

```
71    puts " 第一阶频率 f: [expr $w1/6.28]"
72    set ksi 0.02
73    set a0 0
74    set a1 [expr $ksi*2.0/$w1]
75    rayleigh $a0 0.0 $a1 0.0
76    wipeAnalysis
77    constraints Plain
78    numberer Plain
79    system BandGeneral
80    test NormDispIncr 1.0e-8 10 2
81    algorithm Newton
82    integrator Newmark 0.5 0.25
83    analysis Transient
84    analyze 1000 0.02
85    puts 地面运动分析完成..."
```

(三) 结果分析

(1) 重力作用和水平静力 pushover 作用

整体坐标系下框架的 "pushover 分布力-位移" 曲线：图 1.4.16 中曲线由输出文件 disp_34.out 中第 1 列记录的 pushover 的最大分布力和第 2 列记录的节点 3 的水平位移得到.

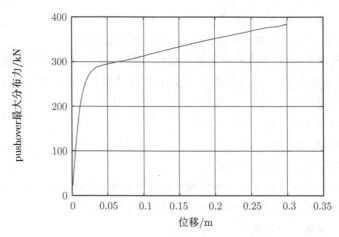

图 1.4.16　pushover 最大分布力-位移

1.4 框架结构分析

局部坐标系下的非弹性混凝土柱 A 点截面的弯矩-曲率曲线: 图 1.4.17 中曲线由输出文件 defocolsec_1.out 中第 3 列记录的第 1 个高斯点在局部坐标系下的曲率和输出文件 forcecolsec_1.out 中第 3 列记录的局部坐标系下的弯矩得到.

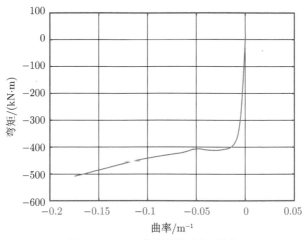

图 1.4.17　A 点截面的弯矩-曲率

(2) 受到最大加速度为 $0.90g(g=9.8\text{m/s}^2)$ 的地震作用

模型水平方向的 "时间-顶点水平位移" 曲线: 图 1.4.18 中的 "顶部节点水平位移" 由输出文件 disp_34.out 中第 1 列记录的时间和第 2 列记录 3 节点的水平方向的位移得到.

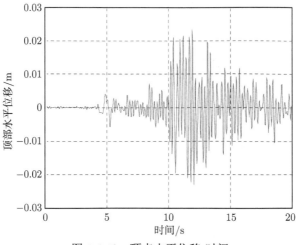

图 1.4.18　顶点水平位移-时间

局部坐标系下的非弹性混凝土柱 A 点截面的弯矩-曲率曲线: 图 1.4.19 中曲线

由输出文件 defocolsec_1.out 中第 3 列记录的第 1 个高斯点在局部坐标系下的曲率和输出文件 forcecolsec_1.out 中第 3 列记录的局部坐标系下的弯矩得到.

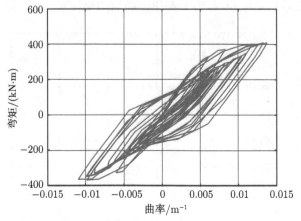

图 1.4.19 A 点截面的弯矩-曲率

1.4.4　三维框架结构在地震下的响应分析

(一) 问题简述

本算例模型为一个三维的三层框架结构建筑模型 (图 1.4.20), 该算例中柱截面为非线性纤维截面, 梁截面为弹性截面, 楼板符合刚性楼板假定. 模型受到水平两个方向地震激励作用, X 方向最大加速度为 $0.900g(g=9.8\mathrm{m/s^2})$, Y 方向最大加速度为 $0.977g$. 以下对建模的 Tcl 输入命令流进行简要的解释:

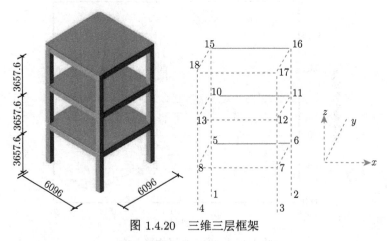

图 1.4.20 三维三层框架

本节算例可参考文献: Barbato M, Gu Q, and Conte J P. Probabilistic Pushover Analysis of Structural and Geotechnical Systems. Journal of Structural Engineering,

(ASCE), 2010, 136, 11: 1330–1341.

(二) 命令流分析

(1) 节点与约束

首先确定算例所用单位为 m、kN、t、s. 定义模型如下:

```
1  model BasicBuilder -ndm 3 -ndf 6
```

第 1 行中定义了模型为三维模型, 每个节点 6 个自由度.

```
2  set h 3.6576
3  set by 6.096
4  set bx 6.096
```

第 2 行到第 4 行定义了变量 h 表示层高, 其数值大小为 3.6576m, 变量 bx 与 by 分别表示结构长与宽, 数值大小都为 6.096m.

```
5   node 1  [expr -$bx/2] [expr $by/2]    0.0
6   node 2  [expr $bx/2]  [expr $by/2]    0.0
7   node 3  [expr $bx/2]  [expr -$by/2]   0.0
8   node 4  [expr -$bx/2] [expr -$by/2]   0.0
9   node 5  [expr -$bx/2] [expr $by/2]    $h
10  node 6  [expr $bx/2]  [expr $by/2]    $h
11  node 7  [expr $bx/2]  [expr -$by/2]   $h
12  node 8  [expr -$bx/2] [expr -$by/2]   $h
13  node 10 [expr -$bx/2] [expr $by/2]    [expr 2*$h]
14  node 11 [expr $bx/2]  [expr $by/2]    [expr 2*$h]
15  node 12 [expr $bx/2]  [expr -$by/2]   [expr 2*$h]
16  node 13 [expr -$bx/2] [expr -$by/2]   [expr 2*$h]
17  node 15 [expr -$bx/2] [expr $by/2]    [expr 3*$h]
18  node 16 [expr $bx/2]  [expr $by/2]    [expr 3*$h]
19  node 17 [expr $bx/2]  [expr -$by/2]   [expr 3*$h]
20  node 18 [expr -$bx/2] [expr -$by/2]   [expr 3*$h]
```

第 5 行到第 20 行定义各个节点的坐标位置.

```
21  node 9  0.0 0.0$h
22  node 14 0.0 0.0 [expr 2*$h]
23  node 19 0.0 0.0[expr 3*$h]
```

第 21 行到第 23 行定义刚性楼板中心处的节点坐标:

```
24  fix 1 1 1 1 1 1 1
25  fix 2 1 1 1 1 1 1
26  fix 3 1 1 1 1 1 1
```

```
27    fix  4  1  1  1  1  1  1
```
第 24 行到第 27 行底部约束条件定义如下：

在三维问题中，上面算例中每个 Tcl 命令都必须拓展到三维中。比如：本算例节点有 6 个自由度，因此命令 fix 命令后需要 6 个值 (而不是以前的 3 个值)："0" 表示自由，"1" 表示固定。算例中四根框架柱底部固支于地面，因此 6 个自由度均固定 (6 个自由度指的是 x 方向，y 方向，z 方向的位移以及绕 x 轴正方向，绕 y 轴正方向，绕 z 轴正方向逆时针的扭转角)。

以下根据四个节点的编号定义刚性楼板，具体命令如下：
```
28    rigidDiaphragm  3   9   5   6   7   8
29    rigidDiaphragm  3  14  10  11  12  13
30    rigidDiaphragm  3  19  15  16  17  18
```
第 28 行到第 30 行的通式可以写为以下形式：rigidDiaphragm $perpDirn $masterNodeTag $slaveNodeTag1 $slaveNodeTag2... 命令中$perpDirn 表示刚性板单元平面的法线方向，如本算例中的 "3" 表示平面法线向量平行于 z 轴方向，即在 xy 平面上的刚性板)，$masterNodeTag 表示主节点编号，$slaveNodeTag1 $slaveNodeTag2... 表示从节点编号。比如本算例中第 28 行表示从节点 5、6、7、8 与主节点 9 的在 xy 平面内为刚性楼板。另外，不考虑刚性板 z 方向位移和绕 x, y 轴转动，因此定义如下约束：
```
31    fix  9  0  0  1  1  1  0
32    fix 14  0  0  1  1  1  0
33    fix 19  0  0  1  1  1  0
```

> **注意**　当模型中有使用 "rigidDiaphragm"，"equalDOF" 等约束的时候，必须用 "constraints Transformation" 或者 "constraints Penalty"，而不能用 "constraints Plain"，当模型中只有 "fix" 约束才可以用 Plain! 具体可以参考有限元教材。当使用 Transformation 时，只保留主节点自由度。从节点自由度被删除，因此再对此节点使用其他约束 (比如 fix) 将出错!

(2) 材料及截面定义

首先定义核心区混凝土材料参数：
```
34    uniaxialMaterial Concrete01 1 -34473.8 -0.005 -24131.66 -0.02
```
定义保护层混凝土材料参数如下：
```
35    set fc 27579.04
36    uniaxialMaterial  Concrete01 2  -$fc -0.002 0.0 -0.006
```
定义钢筋材料参数如下：
```
37    uniaxialMaterial Steel01 3 248200.  2.1e8 0.02
```
定义弹性抗扭材料参数如下

1.4 框架结构分析

```
38   set E 24855585.89304
39   set GJ 68947600000000
40   uniaxialMaterial Elastic 10 $GJ
```

本模型使用非线性纤维截面模型,定义方法与 1.4.3 相同. 为了建模方便,使用 rcsection.tcl 子程序建立纤维截面模型.

```
41   source RCsection.tcl
```

以下介绍子程序中使用纤维模型的建模过程,输入参数主要包括以下:

id-　　程序生成的截面序列号
h -　　截面高
b -　　截面宽
cover - 保护层厚度
coreID - 混凝土核心区分块材料号
coverID - 混凝土保护层分块材料号
steelID - 钢筋材料号
numBars - 截面任何一边的钢筋根数
barArea - 钢筋正截面面积
nfCoreY - 核心区在 Y 方向划分的纤维单元数
nfCoreZ - 核心区在 Z 方向划分的纤维单元数
nfCoverY - 保护层在 Y 方向划分的纤维单元数
nfCoverZ - 保护层在 Z 方向划分的纤维单元数

注意: 在纤维截面定义中,局部坐标系为 Y-Z 平面,三维局部坐标系定义在下面详细讲解.

(3) 程序 RCsection.tcl 的内容如下, 用户可以跳过这节直接后面的学习

```
A1   proc RCsection {id h b cover coreID coverID steelID numBars
barArea nfCoreY nfCoreZ nfCoverY nfCoverZ} {
```

Proc 后为子函数定义, 大括号中表示传入的参数变量名.

```
A2   set coverY [expr $h/2.0] #定义 Y 轴正方向的截面 Z 轴到保护层外
A3   set coverZ [expr $b/2.0] #定义 Z 轴正方向的截面 Y 轴到保护层外
A4   set ncoverY [expr -$coverY] #定义 Y 轴负方向的截面 Z 轴到保护层
A5   set ncoverZ [expr -$coverZ] #定义 Z 轴负方向的截面 Y 轴到保护层
A6   set coreY [expr $coverY-$cover]
A7   set coreZ [expr $coverZ-$cover]
A8   set ncoreY [expr -$coreY]
A9   set ncoreZ [expr -$coreZ]
```

命令 A2 至命令 A9 定义相应保护层厚度、轴线到混凝土核心区外边线的距离等,然后定义纤维截面如下:

```
A10    section fiberSec $id {
```
首先定义核心区混凝土,命令如下:
```
A11    patch quadr $coreID $nfCoreZ $nfCoreY $ncoreY $coreZ $ncoreY
       $ncoreZ $coreY $ncoreZ $coreY $coreZ
```
quadr 表示四边形, $coreID 表示此前定义的截面材料号 (比如材料编号为 1 表示之前定义的混凝土). $nfCoreY、$nfCoreZ 表示截面沿局部坐标 y' 和 z' 坐标轴方向划分的纤维数. 其后的参数表示截面四个顶点坐标,按照逆时针定义. 本例中四个端点的坐标即为内部核心区混凝土的四个顶点的坐标值.

同理将保护层划分为四个四边形,定义纤维截面如下:
```
A12    patch quadr $coverID 1 $nfCoverY $ncoverY $coverZ $ncoreY
       $coreZ $coreY $coreZ $coverY $coverZ
A13    patch quadr $coverID 1 $nfCoverY $ncoreY $ncoverY
       $ncoverZ $coverY $ncoverZ $coreY $ncoreZ
A14    patch quadr $coverID $nfCoverZ 1 $ncoverY $coverZ $ncoverY
       $ncoverZ $ncoreY $ncoreZ $ncoreY $coreZ
A15    patch quadr $coverID $nfCoverZ 1 $coreY $coreZ $coreY $ncoreZ
       $coverY $ncoverZ $coverY $coverZ
```
截面中还有平行于局部坐标 z' 方向钢筋,其具体定义方式如下:
```
A16    layer straight $steelID $numBars $barArea $ncoreY $coreZ
       $ncoreY $ncoreZ
A17    layer straight $steelID $numBars $barArea $coreY $coreZ $coreY
       $ncoreZ
```
命令中 layer straight 表示沿直线分布的钢筋, $steelID 表示钢筋材料号, $numBars 表示钢筋根数, $barArea 表示钢筋面积. $ncoreY、$coreZ 为分布钢筋的起点 y' 和 z' 坐标, $ncoreY、$ncoreZ 表示终点的坐标.

以下计算并用类似方法定义平行于 y' 方向钢筋:
```
A18    set spacingY [expr ($coreY-$ncoreY)/($numBars-1)]
A19    set numBars [expr $numBars-2]
A20    layer straight $steelID $numBars $barArea
       [expr $coreY-$spacingY] $coreZ [expr $ncoreY+$spacingY] $coreZ
A21    layer straight $steelID $numBars $barArea
       [expr $coreY-$spacingY] $ncoreZ [expr $ncoreY+$spacingY] $ncoreZ
}
}
```
最后的两个半个大括号分别表示纤维函数定义结束与子程序的结束.

1.4 框架结构分析

(4) 梁、柱单元定义

基于上述子程序建立纤维截面模型如下第 42 行, 注意各参数含义为:

id h b coverThick coreID coverID steelID nBars area nfCoreY nfCoreZ nfCoverY nfCoverZ

42　RCsection 1 $d $d 0.04 1 2 3 3 0.00051 8 8 10 10

定义了抗弯截面 (1 号截面) 之后, 用 section Aggregator 命令把此截面与抗扭转材料 (10 号材料) 聚合在一起, 定义抗弯扭的截面 (此处没有定义抗剪):

43　section Aggregator 2 10 T -section 1

然后定义单元, 需要先定义坐标转换, 再定义柱单元如下:

44　set colSec 2

45　geomTransf Linear 1 1 0 0

46　set np 4

47　elemenet dispBeamColumn 1 1 5 $np $colSec 1

48　elemenet dispBeamColumn 2 2 6 $np $colSec 1

49　elemenet dispBeamColumn 3 3 7 $np $colSec 1

50　elemenet dispBeamColumn 4 4 8 $np $colSec 1

51　elemenet dispBeamColumn 5 5 10 $np $colSec 1

52　elemenet dispBeamColumn 6 6 11 $np $colSec 1

53　elemenet dispBeamColumn 7 7 12 $np $colSec 1

54　elemenet dispBeamColumn 8 8 13 $np $colSec 1

55　elemenet dispBeamColumn 9 10 15 $np $colSec 1

56　elemenet dispBeamColumn 10 11 16 $np $colSec 1

57　elemenet dispBeamColumn 11 12 17 $np $colSec 1

58　elemenet dispBeamColumn 12 13 18 $np $colSec 1

其中第 45 行用三维的坐标转换, 与上文中二维的坐标转换略有不同, 以下进行详细的介绍.

> **注意**　在 OpenSees 中, 三维的坐标变换命令的格式为
>
> geomTransf Linear $transfTag $vecxzX $vecxzY $vecxzZ
>
> 其中 Linear 指的是线性变换, 即不考虑几何大变形. 后面的参数 $transfTag 表示坐标变换编号. 和二维梁柱模型类似, 局部坐标的 x' 是沿着杆的轴向的, 比如第 47 行定义 1 号单元的局部坐标系 x' 方向是沿杆方向从节点 1 指向节点 5. geomTransf 的后面三个参数 ($vecxzX, $vecxzY, $vecxzZ) 定义了一个新的方向 (比如 (0,0,1) 代表全局坐标的 Z 方向), 此方向为局部坐标系中 $x'z'$ 平面上的一个方向. 将此方向和 x' 方向叉乘 (符合右手规则), 就得到局部坐标的 y' 方向. 然后根据局部坐标的 x' 与 y' 叉乘可得到局部坐标的 z' 方向.

下图为一个一般例子：node i 到 node j 为局部坐标系的 x' 方向 (沿杆方向)，用户定义了 (1,1,0) 向量如图所示. 此向量叉乘局部系 x' 得到局部系 y' 方向，局部系 x' 叉乘局部系 y' 得到局部系 z' 方向，参照图 1.4.21.

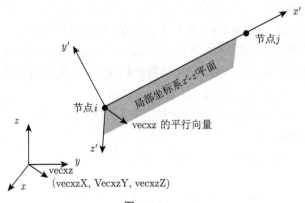

图 1.4.21

以下定义弹性截面的梁单元，截面编号为 3：

59　setAbeam 0.278709
60　setIbeamzz 0.004315
61　setIbeamyy 0.002427
62　section Elastic 3 $E $Abeam $Ibeamzz $Ibeamyy $GJ 1.0

定义坐标转换如下 (具体使用说明见上文)：

63　set beamSec 3
64　geomTransfLinear 2 1 1 0

定义梁单元如下 (具体使用方法同上文)：

65　set np 3
66　element dispBeamColumn 13 5 6 $np $beamSec 2
67　element dispBeamColumn 14 6 7 $np $beamSec 2
68　element dispBeamColumn 15 7 8 $np $beamSec 2
69　element dispBeamColumn 16 8 7 $np $beamSec 2
70　element dispBeamColumn 17 10 11 $np $beamSec 2
71　element dispBeamColumn 18 11 12 $np $beamSec 2
72　element dispBeamColumn 19 12 13 $np $beamSec 2
73　element dispBeamColumn 20 13 10$np $beamSec 2
74　element dispBeamColumn 21 15 16 $np $beamSec 2
75　element dispBeamColumn 22 16 17 $np $beamSec 2

```
76    element dispBeamColumn 23 17 18 $np $beamSec 2
77    element dispBeamColumn 24 18 15 $np $beamSec 2
```

本算例中的两个单元 1(左柱) 和单元 13(梁) 的局部坐标如图 1.4.22 所示.

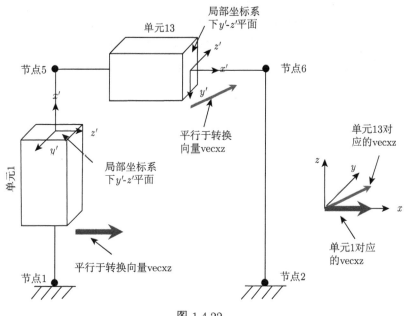

图 1.4.22

(5) 质量定义

将质量分布于板的主节点上:

```
78    set g 9.8;
79    set m 30.0;
80    set i [expr $m*($bx*$bx+$by*$by)/12.0]
81    mass 9 $m $m 0 0 0 $i
82    mass 14 $m $m 0 0 0 $i
83    mass 19 $m $m 0 0 0 $i
```

命令中 mass 后第一个表示节点编号, 随后 6 个数值分别表示 x 轴方向, y 轴方向, z 轴方向质量以及绕 x 轴方向, y 轴方向, z 轴方向旋转时的转动惯量.

(6) 荷载定义

首先对结构施加重力荷载, 使用静力加载的方式, 具体命令如下:

```
84    set p 74.0
85    pattern Plain 1 {Series -time {0.0 2.0 10000.0} -values {0.0 1.0
      1.0}}{foreach node {5 6 7 8 10 11 12 13 15 16 17 18} {load $node
```

```
                0.0 0.0 -$p 0.0 0.0 0.0 }}
```

具体命令细节说明如前文所述,本算例中通过 Series 命令定义时间序列. 具体是通过 -time 命令以及 -values 命令定义曲线如图 1.4.23. 节点 5、6、7 等在系统时间为 0 秒时受力为 0,在 2.0 秒到 10000.0 秒时间内受到 z 方向的力,大小为 -$p(即方向为 z 的反方向). 在每个分析步的系统时间确定后,OpenSees 自动根据图 1.4.23 插值得到此时刻的外力,加载系统上 (比如 3.01 秒时,作用在 5、6、7 等节点上的外力根据插值得到,为 z 方向的力,大小为 -$p).

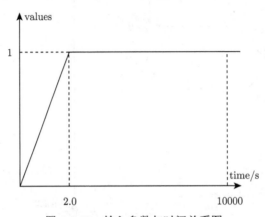

图 1.4.23 输入参数与时间关系图

> **注意** OpenSees 中的系统时间总是有很小的误差 (由于 C++ 语言的局限导致),比如我们希望用固定时间步长 0.01 秒分析时,而实际上系统时间可能是 0.0099999999999 或者 0.0100000000001,这样的小误差本来不会造成错误 (插值后得到的外力值误差很小),但是在最后一步,比如图 1.4.23 中的 10000.0 秒时刻分析可能出错! 如果这一步系统真实时间为 10000.00000000001,由于超出了 pattern 1 所定义的时间长度,系统将不再插值,而直接返回外力为 0! 这时候会导致此时步分析时的错误!

下面添加两个方向的地震:

```
86  set tabasFN "Path -filePath tabasFN.txt -dt 0.02 -factor $g"
87  set tabasFP "Path -filePath tabasFP.txt -dt 0.02 -factor $g"
```

表示导入加速度文件 tabasFN.txt 与 tabasFP.txt,加速度文件中数据的时间间隔为 0.02s,文件中数据必须乘以$g 作为加速度值大小. 输入基底激励外荷载:

```
88  pattern UniformExcitation 2 1 -accel $tabasFN
89  pattern UniformExcitation 3 2 -accel $tabasFP
```

记录的命令如下:

1.4 框架结构分析

```
90   recorder Node -file node.out -time -node 9 14 19 -dof 1 2 3 4 5
     6 disp
```

表示记录 9,14,19 号节点 (板中心) 的 6 个自由度方向的位移响应. 先做重力分析 (静力分析):

```
91   constraints Transformation  # 此处不能用 Plain, 因为是多点约束
92   test EnergyIncr 1.0e-16 20 2
93   integrator LoadControl 1 1 1 1
94   algorithm Newton
95   system BandGeneral
96   numberer RCM
97   analysis Static
98   set startT [clock seconds]    #表示开始记录计算时间
99   analyze 3
```

注意 此时系统时间变为 3.0 秒, 其后的重力 (参考图 1.4.23) 不再改变, 因此可以不用 loadConst -time 0 命令.

此后进行地震动力分析:

```
100  wipe Analysis
101  test EnergyIncr 1.0e-16 20 2
102  algorithm Newton
103  system BandGeneral
104  constraints Transformation
105  numberer RCM
106  integrator Newmark 0.55 0.275625
107  analysis Transient
108  analyze 2500 0.01
109  set endT [clock seconds]
110  puts " 完成时间:  [expr $endT-$startT] seconds."
```

使用牛顿迭代法进行分析, 其中 Newmark 法的参数 α 与 β 分别为 0.55, 0.275625. 通常情况下参数 α 与 β 应该取 0.5 和 0.25, 无条件稳定. **但是有时为了增加非线性体系的收敛性, 也可用其他值, 本算例会导致数值阻尼**(具体参考结构动力学). [clock seconds] 记录当前时间, 用于确定分析所用时间.

> **注意** 由于在静、动力分析之间由于没有使用 loadConst -time 0 命令, 在动力分析初始时刻系统时间就是 3.0 秒了. 因此在 tabasFN.txt 和 tabasFP.txt 中需要在有效记录的前面各添加 3.0 秒空闲时间记录. 由于这两个文件的时间间隔为 0.02 秒, 故各加 150 个 0.0 即可.
>
> **注意** 这些操作完全可以用 loadConst -time 0 命令代替, 这个算例采用这些繁琐做法是为了说明系统时间的概念, 用户实际使用中可不必这样做. 仅在某些特殊情况下才有必要这么做 (比如敏感性分析时必须这样做, 因为敏感性分析算法和 loadConst -time 0 命令不兼容).

(三) 结果分析

如图 1.4.24 所示, 三条不同颜色的线条分别表示第一、二、三层 x 方向层间位移时程响应. 可知结果中楼层 x 方向最大层间位移出现在 11.09s 时, 第一层楼板中心位置, 数值大小为 0.0769m, 第二层的最大层间位移出现在 11.03s, 数值大小为 0.0508m, 第三层最大层间位移出现在 10.99s, 数值大小为 0.0355m.

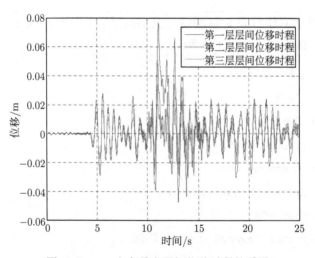

图 1.4.24 x 方向最大层间位移时程关系图

以下为 y 方向最大层间位移时程关系图:

如图 1.4.25 所示, 三条不同颜色的线条分别表示第一、二、三层 y 方向层间位移时程响应. 可知结果中楼层 x 方向最大层间位移出现在 10.56s 时, 第一层楼板中心位置, 数值大小为 0.0953m, 第二层的最大层间位移出现在 10.50s, 数值大小为 0.0545m, 第三层最大层间位移出现在 10.47s, 数值大小为 0.0281m.

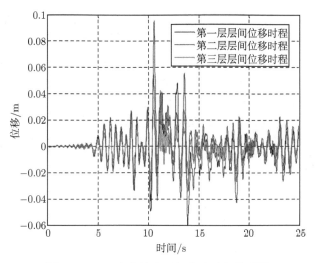

图 1.4.25　y 方向最大层间位移时程关系图

1.5　土-结构相互作用体系

两层框架结构与地基土的相互作用体系地震动力分析算例

(一) 问题简述

> **注意**　土节点 dof=2(无转动), 梁柱节点 dof=3, 因此必须分别建模. 本例中将土与梁柱单元在相同位置处的节点强制粘接在一起, 未考虑相对滑移, 为简化的处理方法. 在不计算敏感性分析时, quadWithSensitivity 与 quad 相同.

本算例为模拟两层框架结构在地震荷载作用下与地基土的相互作用, 模型如图 1.5.1 所示. 框架结构的梁柱截面为非弹性纤维截面, 中柱截面大小为 $0.6\text{m} \times 0.5\text{m}$,

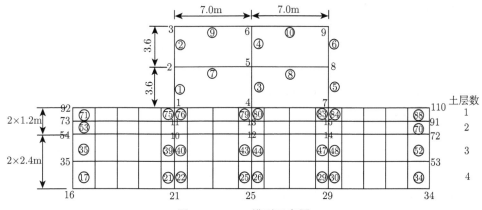

图 1.5.1　SSI 模型示意图

边柱截面为 0.5m×0.5m, 梁截面大小为 0.5m×0.4m. 基础采用弹性混凝土材料模拟, 土使用多屈服面粘土塑性模型模拟, 从上到下采用四种不同材料参数, 模拟土的性质随高度的变化. 土的边界采用简化剪切边界条件. 土与结构体系的底部受到大小为 El_Centro 地震 3 倍的基底一致激励, 最大加速度为 8.769m/s^2 的.

图中给出上部框架结构、基础以及部分土的节点、单元编号 (加圆圈的数字), 方便读者与代码对照. 由于上部框架结构和地基土的自由度不同 (节点自由度分别是 3、2), 需要分开建模. 先施加重力进行静力分析, 再施加地震荷载进行动力分析.

本算例所提到的模型和计算方法可参考文献:

Barbato M, Gu Q*, and Conte J P. 2010. Probabilistic Pushover Analysis of Structural and Geotechnical Systems. Journal of Structural Engineering, (ASCE), 136(11): 1330-1341.

(二) 命令流分析

第一步建立上部框架结构, 为二维, 自由度为 3(节点位移为 x、y 方向的平动和绕 z 轴的转动).

```
1   wipe;
2   model BasicBuilder -ndm2 -ndf3
3   if { [file exists output] == 0 } {
4     file mkdir output;
    }
5   set framemass1    15.0
6   set framemass2    30.0
7   set framemass3    4.0
8   node  1   0.0  0.0   -mass  $framemass1  $framemass1  0.0
9   node  2   0.0  3.6   -mass  $framemass1  $framemass1  0.0
10  node  3   0.0  7.2   -mass  $framemass1  $framemass1  0.0
11  node  4   7.0  0.0   -mass  $framemass2  $framemass2  0.0
12  node  5   7.0  3.6   -mass  $framemass2  $framemass2  0.0
13  node  6   7.0  7.2   -mass  $framemass2  $framemass2  0.0
14  node  7  14.0  0.0   -mass  $framemass1  $framemass1  0.0
15  node  8  14.0  3.6   -mass  $framemass1  $framemass1  0.0
16  node  9  14.0  7.2   -mass  $framemass1  $framemass1  0.0
17  node 10   0.0 -2.4   -mass  $framemass3  $framemass3  0.0
18  node 11   0.0 -1.2   -mass  $framemass3  $framemass3  0.0
19  node 12   7.0 -2.4   -mass  $framemass3  $framemass3  0.0
20  node 13   7.0 -1.2   -mass  $framemass3  $framemass3  0.0
```

```
21  node 14 14.0 -2.4  -mass $framemass3 $framemass3 0.0
22  node 15 14.0 -1.2  -mass $framemass3 $framemass3 0.0
```
以上代码为模型维数、自由度；节点标号、质量信息.
```
23  recorder Node -file output/disp6.out -time -node 6 -dof 1 2 disp
24  recorder Node -file output/disp5.out -time -node 5 -dof 1 2 disp
25  recorder Node -file output/disp4.out -time -node 4 -dof 1 2 disp
```
以上代码记录了输出的节点位移.
```
26  set upperload1 [expr -$framemass1*10.0]
27  set upperload2 [expr -$framemass2*10.0]
28  set download3 [expr -$framemass3*10.0]
29  pattern Plain 2 "Constant" {
30   load 1 0.0 $upperload1 0.0
31   load 2 0.0 $upperload1 0.0
32   load 3 0.0 $upperload1 0.0
33   load 4 0.0 $upperload2 0.0
34   load 5 0.0 $upperload2 0.0
35   load 6 0.0 $upperload2 0.0
36   load 7 0.0 $upperload1 0.0
37   load 8 0.0 $upperload1 0.0
38   load 9 0.0 $upperload1 0.0
39   load 10 0.0 $download3 0.0
40   load 11 0.0 $download3 0.0
41   load 12 0.0 $download3 0.0
42   load 13 0.0 $download3 0.0
43   load 14 0.0 $download3 0.0
44   load 15 0.0 $download3 0.0
45  }
```
以上代码表示框架结构的重力集中在各个节点上.
```
46  uniaxialMaterial Concrete01 1 -27588.5 -0.002 0.0 -0.008
47  uniaxialMaterial Concrete01 2 -34485.6 -0.004 -20691.4 -0.014
48  uniaxialMaterial Hardening  3 2.0e8 248200.0 0.0 1.6129e6
49  uniaxialMaterial Concrete01 4 -27588.5 -0.002 0.0 -0.008
50  uniaxialMaterial Concrete01 5 -34485.6 -0.004 -20691.4 -0.014
51  uniaxialMaterial Hardening  6 2.0e8 248200.0 0.0 1.6129e6
```
以上代码为材料信息. 框架梁柱的混凝土采用 concrete01, 钢筋采用 Hardening.

对于第 46、47、49、50 行代码，每行数字依次代表材料标号、极限强度、达到极限强度对应的应变、混凝土被压碎对应强度、压碎时对应的应变. 中柱、边柱混凝土保护层都用标号为 1 的混凝土，中柱、边柱混凝土核心区用标号为 2 的混凝土材料. 梁用标号为 1 的混凝土. 注意：这里梁没有保护层.

第 48、51 行分别描述中柱、边柱的配筋，每行数字依次代表材料标号、钢筋弹性模量、屈服强度、各向同性强化模量、运动强化模量.

```
52   section Fiber 1 {
53     patch quad 2 1 12 -0.2500  0.2000 -0.2500 -0.2000  0.2500
       -0.2000  0.2500  0.2000
54     patch quad 1 1 14 -0.3000 -0.2000 -0.3000 -0.2500  0.3000
       -0.2500  0.3000 -0.2000
55     patch quad 1 1 14 -0.3000  0.2500 -0.3000  0.2000  0.3000
        0.2000  0.3000  0.2500
56     patch quad 1 1  2 -0.3000  0.2000 -0.3000 -0.2000 -0.2500
       -0.2000 -0.2500  0.2000
57     patch quad 1 1  2  0.2500  0.2000  0.2500 -0.2000  0.3000
       -0.2000  0.3000  0.2000
58     layer straight 3 3 0.000645 -0.2000  0.2000 -0.2000 -0.2000
59     layer straight 3 3 0.000645  0.2000  0.2000  0.2000 -0.2000
60   }
61   section Fiber 2 {
62     patch quad 2 1 10 -0.2000  0.2000 -0.2000 -0.2000  0.2000
       -0.2000  0.2000  0.2000
63     patch quad 1 1 12 -0.2500 -0.2000 -0.2500 -0.2500  0.2500
       -0.2500  0.2500 -0.2000
64     patch quad 1 1 12 -0.2500  0.2500 -0.2500  0.2000  0.2500
        0.2000  0.2500  0.2500
65     patch quad 1 1  2 -0.2500  0.2000 -0.2500 -0.2000 -0.2000
       -0.2000 -0.2000  0.2000
66     patch quad 1 1  2  0.2000  0.2000  0.2000 -0.2000  0.2500
       -0.2000  0.2500  0.2000
67     layer straight 3 3 0.00051 -0.2000  0.2000 -0.2000 -0.2000
68     layer straight 3 3 0.00051  0.2000  0.2000  0.2000 -0.2000}
69   section Fiber 3 {
```

```
70  patch quad 1 1 12 -0.2500  0.2000 -0.2500 -0.2000  0.2500
    -0.2000  0.2500  0.2000
71  layer straight 3 2 0.000645 -0.2000  0.2000 -0.2000 -0.2000
72  layer straight 3 2 0.000645  0.2000  0.2000  0.2000 -0.2000
    }
73  section Fiber 4 {
74  patch quad 5 1 12 -0.2500  0.2000 -0.2500 -0.2000  0.2500
    -0.2000  0.2500  0.2000
75  patch quad 4 1 14 -0.3000 -0.2000 -0.3000 -0.2500  0.3000
    -0.2500  0.3000 -0.2000
76  patch quad 4 1 14 -0.3000  0.2500 -0.3000  0.2000  0.3000
    0.2000  0.3000  0.2500
77  patch quad 4 1 2 -0.3000  0.2000 -0.3000 -0.2000 -0.2500
    -0.2000 -0.2500  0.2000
78  patch quad 4 1 2  0.2500  0.2000  0.2500 -0.2000  0.3000
    -0.2000  0.3000  0.2000
79  layer straight 6 3 0.000645 -0.2000  0.2000 -0.2000 -0.2000
80  layer straight 6 3 0.000645  0.2000  0.2000  0.2000 -0.2000
81  }
82  section Fiber 5 {
83  patch quad 5 1 10 -0.2000  0.2000 -0.2000 -0.2000  0.2000
    -0.2000  0.2000  0.2000
84  patch quad 4 1 12 -0.2500 -0.2000 -0.2500 -0.2500  0.2500
    -0.2500  0.2500 -0.2000
85  patch quad 4 1 12 -0.2500  0.2500 -0.2000  0.2000  0.2500
    0.2000  0.2500  0.2500
86  patch quad 4 1 2 -0.2500  0.2000 -0.2500 -0.2000 -0.2000
    -0.2000 -0.2000  0.2000
87  patch quad 4 1 2  0.2000  0.2000  0.2000 -0.2000  0.2500
    -0.2000  0.2500  0.2000
88  layer straight 6 3 0.00051 -0.2000  0.2000 -0.2000 -0.2000
89  layer straight 6 3 0.00051  0.2000  0.2000  0.2000 -0.2000
    }
```

以上代码描述构件的截面，包括上部结构的梁和柱、下部基础，全部采用纤维截面．

```
90    set nP 4
91    geomTransf Linear   1
92    element dispBeamColumn   1    1    2   $nP   2   1
93    element dispBeamColumn   2    2    3   $nP   2   1
94    element dispBeamColumn   3    4    5   $nP   1   1
95    element dispBeamColumn   4    5    6   $nP   1   1
96    element dispBeamColumn   5    7    8   $nP   2   1
97    element dispBeamColumn   6    8    9   $nP   2   1
98    element dispBeamColumn   7    2    5   $nP   3   1
99    element dispBeamColumn   8    5    8   $nP   3   1
100   element dispBeamColumn   9    3    6   $nP   3   1
101   element dispBeamColumn   10   6    9   $nP   3   1
102   element dispBeamColumn   11   10   11  $nP   4   1
103   element dispBeamColumn   12   11   1   $nP   4   1
104   element dispBeamColumn   13   12   13  $nP   5   1
105   element dispBeamColumn   14   13   4   $nP   5   1
106   element dispBeamColumn   15   14   15  $nP   4   1
107   element dispBeamColumn   16   15   7   $nP   4   1
```

第 92-97 行是柱单元，第 98-101 行是梁单元，第 102-107 行是下部基础单元的一部分。事实上，11、12 号梁柱单元与下部结构的第 57、58、75、76 号土单元，13、14 号梁柱单元与下部结构的第 61、62、79、80 号土单元，11、12 号梁柱单元与下部结构的第 65、66、83、84 号土单元，分别构成结构的 3 个基础。以上全部代码是上部框架结构的模型。

```
108   recorder Element -ele 1 2 -file output/Deformation12.out -time
      section 2 deformations
109   recorder Element -ele 1 2 -file output/Force12.out -time section
      2 force
110   recorder Element -ele 3 4 -file output/Deformation34.out -time
      section 2 deformations
111   recorder Element -ele 3 4 -file output/Force34.out -time section
      2 force
112   recorder Element -ele 7 9 -file output/Deformation79.out -time
      section 3 deformations
113   recorder Element -ele 7 9 -file output/Force79.out -time section
      3 force
```

```
114  recorder Element -ele 7 -time -file output/steelstress7.out
     section 3 fiber -0.2286 0.2286 stress
115  recorder Element -ele 7 -time -file output/steelstrain7.out
     section 3 fiber -0.2286 0.2286 strain
116  recorder Element -ele 7 -time -file output/concretestress7.out
     section 3 fiber 0.0 0.0 stress
117  recorder Element -ele 7 -time -file output/concretestrain7.out
     section 3 fiber 0.0 0.0 strain
```
以上第 108-117 行代码是记录输入内容.

**第二步是下部地基土模型, 为二维, 节点自由度为 $2(x、y$ 两个方向的位移).
模型如下:**

```
118  set g -19.6
119  model basic -ndm 2 -ndf 2
120  node 16   -9.2  -7.2
121  node 17   -7.2  -7.2
122  node 18   -5.2  -7.2
123  node 19   -3.2  -7.2
124  node 20   -1.2  -7.2
125  node 21    0.0  -7.2
126  node 22    1.2  -7.2
127  node 23    3.5  -7.2
128  node 24    5.8  -7.2
129  node 25    7.0  -7.2
130  node 26    8.2  -7.2
131  node 27   10.5  -7.2
132  node 28   12.8  -7.2
133  node 29   14.0  -7.2
134  node 30   15.2  -7.2
135  node 31   17.2  -7.2
136  node 32   19.2  -7.2
137  node 33   21.2  -7.2
138  node 34   23.2  -7.2
139  node 35   -9.2  -4.8
140  node 36   -7.2  -4.8
141  node 37   -5.2  -4.8
```

```
142    node 38    -3.2   -4.8
143    node 39    -1.2   -4.8
144    node 40     0.0   -4.8
145    node 41     1.2   -4.8
146    node 42     3.5   -4.8
147    node 43     5.8   -4.8
148    node 44     7.0   -4.8
149    node 45     8.2   -4.8
150    node 46    10.5   -4.8
151    node 47    12.8   -4.8
152    node 48    14.0   -4.8
153    node 49    15.2   -4.8
154    node 50    17.2   -4.8
155    node 51    19.2   -4.8
156    node 52    21.2   -4.8
157    node 53    23.2   -4.8
158    node 54    -9.2   -2.4
159    node 55    -7.2   -2.4
160    node 56    -5.2   -2.4
161    node 57    -3.2   -2.4
162    node 58    -1.2   -2.4
163    node 59     0.0   -2.4
164    node 60     1.2   -2.4
165    node 61     3.5   -2.4
166    node 62     5.8   -2.4
167    node 63     7.0   -2.4
168    node 64     8.2   -2.4
169    node 65    10.5   -2.4
170    node 66    12.8   -2.4
171    node 67    14.0   -2.4
172    node 68    15.2   -2.4
173    node 69    17.2   -2.4
174    node 70    19.2   -2.4
175    node 71    21.2   -2.4
176    node 72    23.2   -2.4
```

177	node 73	-9.2	-1.2
178	node 74	-7.2	-1.2
179	node 75	-5.2	-1.2
180	node 76	-3.2	-1.2
181	node 77	-1.2	-1.2
182	node 78	0.0	-1.2
183	node 79	1.2	-1.2
184	node 80	3.5	-1.2
185	node 81	5.8	-1.2
186	node 82	7.0	-1.2
187	node 83	8.2	-1.2
188	node 84	10.5	-1.2
189	node 85	12.8	-1.2
190	node 86	14.0	-1.2
191	node 87	15.2	-1.2
192	node 88	17.2	-1.2
193	node 89	19.2	-1.2
194	node 90	21.2	-1.2
195	node 91	23.2	-1.2
196	node 92	-9.2	0.0
197	node 93	-7.2	0.0
198	node 94	-5.2	0.0
199	node 95	-3.2	0.0
200	node 96	-1.2	0.0
201	node 97	0.0	0.0
202	node 98	1.2	0.0
203	node 99	3.5	0.0
204	node 100	5.8	0.0
205	node 101	7.0	0.0
206	node 102	8.2	0.0
207	node 103	10.5	0.0
208	node 104	12.8	0.0
209	node 105	14.0	0.0
210	node 106	15.2	0.0
211	node 107	17.2	0.0

```
212  node 108  19.2  0.0
213  node 109  21.2  0.0
214  node 110  23.2  0.0
```

第 120-214 行表示土节点.

```
215  fix 16 1 1
216  fix 17 1 1
217  fix 18 1 1
218  fix 19 1 1
219  fix 20 1 1
220  fix 21 1 1
221  fix 22 1 1
222  fix 23 1 1
223  fix 24 1 1
224  fix 25 1 1
225  fix 26 1 1
226  fix 27 1 1
227  fix 28 1 1
228  fix 29 1 1
229  fix 30 1 1
230  fix 31 1 1
231  fix 32 1 1
232  fix 33 1 1
233  fix 34 1 1
```

215-233 行定义地基土的边界条件, 土最底部节点的 x、y 方向是固定的.

```
234  nDMaterial  MultiYieldSurfaceClay 101 2 0.0 54450 1.6e5
     33.00.1
235  nDMaterial  MultiYieldSurfaceClay 102 2 0.0 33800 1.0e5
     26.00.1
236  nDMaterial  MultiYieldSurfaceClay 103 2 0.0 61250 1.8e5
     35.00.1
237  nDMaterial  MultiYieldSurfaceClay 104 2 0.0 96800 2.9e5
     44.00.1
238  nDMaterial  MultiYieldSurfaceClay 100 2 0.0  2e7  1.0e6
     21000. 50.0
```

第 234-237 行是土的材料, 为多屈服面粘土. 后面数字分别表示材料标号、计算维数、土的质量密度 (材料密度为 0, 单元密度非 0)、剪切模量、体积模量、屈服强度、最大切应变. 可以看到, 四种土材料参数都不同, 分别代表地基的四层不同土.

第 238 行是基础的材料, 屈服强度和最大切应变的数值都取很大, 保证在受荷载作用下保持弹性.

```
239 element quadWithSensitivity 17 16 17 36 35 0.60 "PlaneStrain"
    104 0 2.0 0 $g
240 element quadWithSensitivity 18 17 18 37 36 0.60 "PlaneStrain"
    104 0 2.0 0 $g
241 element quadWithSensitivity 19 18 19 38 37 0.60 "PlaneStrain"
    104 0 2.0 0 $g
242 element quadWithSensitivity 20 19 20 39 38 0.60 "PlaneStrain"
    104 0 2.0 0 $g
243 element quadWithSensitivity 21 20 21 40 39 0.60 "PlaneStrain"
    104 0 2.0 0 $g
244 element quadWithSensitivity 22 21 22 41 40 0.60 "PlaneStrain"
    104 0 2.0 0 $g
245 element quadWithSensitivity 23 22 23 42 41 0.60 "PlaneStrain"
    104 0 2.0 0 $g
246 element quadWithSensitivity 24 23 24 43 42 0.60 "PlaneStrain"
    104 0 2.0 0 $g
247 element quadWithSensitivity 25 24 25 44 43 0.60 "PlaneStrain"
    104 0 2.0 0 $g
248 element quadWithSensitivity 26 25 26 45 44 0.60 "PlaneStrain"
    104 0 2.0 0 $g
249 element quadWithSensitivity 27 26 27 46 45 0.60 "PlaneStrain"
    104 0 2.0 0 $g
250 element quadWithSensitivity 28 27 28 47 46 0.60 "PlaneStrain"
    104 0 2.0 0 $g
251 element quadWithSensitivity 29 28 29 48 47 0.60 "PlaneStrain"
    104 0 2.0 0 $g
252 element quadWithSensitivity 30 29 30 49 48 0.60 "PlaneStrain"
    104 0 2.0 0 $g
```

```
253 element quadWithSensitivity 31 30 31 50 49 0.60 "PlaneStrain"
    104 0 2.0 0 $g
254 element quadWithSensitivity 32 31 32 51 50 0.60 "PlaneStrain"
    104 0 2.0 0 $g
255 element quadWithSensitivity 33 32 33 52 51 0.60 "PlaneStrain"
    104 0 2.0 0 $g
256 element quadWithSensitivity 34 33 34 53 52 0.60 "PlaneStrain"
    104 0 2.0 0 $g
257 element quadWithSensitivity 35 35 36 55 54 0.60 "PlaneStrain"
    103 0 2.0 0 $g
258 element quadWithSensitivity 36 36 37 56 55 0.60 "PlaneStrain"
    103 0 2.0 0 $g
259 element quadWithSensitivity 37 37 38 57 56 0.60 "PlaneStrain"
    103 0 2.0 0 $g
260 element quadWithSensitivity 38 38 39 58 57 0.60 "PlaneStrain"
    103 0 2.0 0 $g
261 element quadWithSensitivity 39 39 40 59 58 0.60 "PlaneStrain"
    103 0 2.0 0 $g
262 element quadWithSensitivity 40 40 41 60 59 0.60 "PlaneStrain"
    103 0 2.0 0 $g
263 element quadWithSensitivity 41 41 42 61 60 0.60 "PlaneStrain"
    103 0 2.0 0 $g
264 element quadWithSensitivity 42 42 43 62 61 0.60 "PlaneStrain"
    103 0 2.0 0 $g
265 element quadWithSensitivity 43 43 44 63 62 0.60 "PlaneStrain"
    103 0 2.0 0 $g
266 element quadWithSensitivity 44 44 45 64 63 0.60 "PlaneStrain"
    103 0 2.0 0 $g
267 element quadWithSensitivity 45 45 46 65 64 0.60 "PlaneStrain"
    103 0 2.0 0 $g
268 element quadWithSensitivity 46 46 47 66 65 0.60 "PlaneStrain"
    103 0 2.0 0 $g
269 element quadWithSensitivity 47 47 48 67 66 0.60 "PlaneStrain"
    103 0 2.0 0 $g
```

```
270 element quadWithSensitivity 48 48 49 68 67 0.60 "PlaneStrain"
    103 0 2.0 0 $g
271 element quadWithSensitivity 49 49 50 69 68 0.60 "PlaneStrain"
    103 0 2.0 0 $g
272 element quadWithSensitivity 50 50 51 70 69 0.60 "PlaneStrain"
    103 0 2.0 0 $g
273 element quadWithSensitivity 51 51 52 71 70 0.60 "PlaneStrain"
    103 0 2.0 0 $g
274 element quadWithSensitivity 52 52 53 72 71 0.60 "PlaneStrain"
    103 0 2.0 0 $g
275 element quadWithSensitivity 53 54 55 74 73 0.60 "PlaneStrain"
    102 0 2.0 0 $g
276 element quadWithSensitivity 54 55 56 75 74 0.60 "PlaneStrain"
    102 0 2.0 0 $g
277 element quadWithSensitivity 55 56 57 76 75 0.60 "PlaneStrain"
    102 0 2.0 0 $g
278 element quadWithSensitivity 56 57 58 77 76 0.60 "PlaneStrain"
    102 0 2.0 0 $g
279 element quadWithSensitivity 57 58 59 78 77 0.60 "PlaneStrain"
    100 0 2.0 0 $g
280 element quadWithSensitivity 58 59 60 79 78 0.60 "PlaneStrain"
    100 0 2.0 0 $g
281 element quadWithSensitivity 59 60 61 80 79 0.60 "PlaneStrain"
    102 0 2.0 0 $g
282 element quadWithSensitivity 60 61 62 81 80 0.60 "PlaneStrain"
    102 0 2.0 0 $g
283 element quadWithSensitivity 61 62 63 82 81 0.60 "PlaneStrain"
    100 0 2.0 0 $g
284 element quadWithSensitivity 62 63 64 83 82 0.60 "PlaneStrain"
    100 0 2.0 0 $g
285 element quadWithSensitivity 63 64 65 84 83 0.60 "PlaneStrain"
    102 0 2.0 0 $g
286 element quadWithSensitivity 64 65 66 85 84 0.60 "PlaneStrain"
    102 0 2.0 0 $g
```

287 element quadWithSensitivity 65 66 67 86 85 0.60 "PlaneStrain" 100 0 2.0 0 $g
288 element quadWithSensitivity 66 67 68 87 86 0.60 "PlaneStrain" 100 0 2.0 0 $g
289 element quadWithSensitivity 67 68 69 88 87 0.60 "PlaneStrain" 102 0 2.0 0 $g
290 element quadWithSensitivity 68 69 70 89 88 0.60 "PlaneStrain" 102 0 2.0 0 $g
291 element quadWithSensitivity 69 70 71 90 89 0.60 "PlaneStrain" 102 0 2.0 0 $g
292 element quadWithSensitivity 70 71 72 91 90 0.60 "PlaneStrain" 102 0 2.0 0 $g
293 element quadWithSensitivity 71 73 74 93 92 0.60 "PlaneStrain" 101 0 2.0 0 $g
294 element quadWithSensitivity 72 74 75 94 93 0.60 "PlaneStrain" 101 0 2.0 0 $g
295 element quadWithSensitivity 73 75 76 95 94 0.60 "PlaneStrain" 101 0 2.0 0 $g
296 element quadWithSensitivity 74 76 77 96 95 0.60 "PlaneStrain" 101 0 2.0 0 $g
297 element quadWithSensitivity 75 77 78 97 96 0.60 "PlaneStrain" 100 0 2.0 0 $g
298 element quadWithSensitivity 76 78 79 98 97 0.60 "PlaneStrain" 100 0 2.0 0 $g
299 element quadWithSensitivity 77 79 80 99 98 0.60 "PlaneStrain" 101 0 2.0 0 $g
300 element quadWithSensitivity 78 80 81 100 99 0.60 "PlaneStrain" 101 0 2.0 0 $g
301 element quadWithSensitivity 79 81 82 101 100 0.60 "PlaneStrain" 100 0 2.0 0 $g
302 element quadWithSensitivity 80 82 83 102 101 0.60 "PlaneStrain" 100 0 2.0 0 $g
303 element quadWithSensitivity 81 83 84 103 102 0.60 "PlaneStrain" 101 0 2.0 0 $g

1.5 土-结构相互作用体系

```
304 element quadWithSensitivity 82 84 85 104 103 0.60 "PlaneStrain"
    101 0 2.0 0 $g
305 element quadWithSensitivity 83 85 86 105 104 0.60 "PlaneStrain"
    100 0 2.0 0 $g
306 element quadWithSensitivity 84 86 87 106 105 0.60 "PlaneStrain"
    100 0 2.0 0 $g
307 element quadWithSensitivity 85 87 88 107 106 0.60 "PlaneStrain"
    101 0 2.0 0 $g
308 element quadWithSensitivity 86 88 89 108 107 0.60 "PlaneStrain"
    101 0 2.0 0 $g
309 element quadWithSensitivity 87 89 90 109 108 0.60 "PlaneStrain"
    101 0 2.0 0 $g
310 element quadWithSensitivity 88 90 91 110 109 0.60 "PlaneStrain"
    101 0 2.0 0 $g
```

采用 quadWithSensitivity 平面四节点单元, 与 quad 平面四节点单元类似.

按顺序解释每行代码的含义, 例如第 310 行, 88 为土单元标号, 该单元的四个节点标号分别是 90、91、110、109 (注意: 四个节点必须是逆时针顺序). 0.6 是单元厚度 (沿 z 方向), "PlaneStrain" 表示平面应变, 使用标号为 101 的材料, 单元表面压力为 0. 2.0 是单元的质量密度, 最后两个数字分别是 x、y 方向的体力.

第 57、58、75、76、61、62、79、80、65、66、83、84 号单元, 采用标号为 100 的材料, 与上部梁柱结构构成基础 (上面已有提到). 除此之外, 第 240-256 行是土的第四层单元, 第 257-274 行是土的第三层单元, 第 275-292 行是土的第二层单元, 第 293-310 行是土的第一层单元.

> **注意** 土材料要输入密度, 土单元也要输入密度, 在 OpenSees 内部计算时, 把两个密度相加, 所以这里把多屈服面粘土材料密度设为 0.0, 把四节点单元密度设为 2.0(吨/立方米) 就可以了.

第三步是设定边界条件. 首先设定土的左右两边界必须保持位移同步, 模拟简化的剪切边界条件, 即认为所取的土在地震作用下做简单剪切运动.

```
311    equalDOF 16   34 1 2;  由于 16 和 34 均 fix, 此行不用加
312    equalDOF 35   53 1 2
313    equalDOF 54   72 1 2
314    equalDOF 73   91 1 2
315    equalDOF 92  110 1 2
```

其次通过边界条件把以上定义的结构和土连接起来, 即把桩和土相同位置处

的一对节点的平动位移粘接在一起。这是一种简单做法,如果需要考虑桩土间的滑移、桩的直径、以及土不能抗拉等效应,用户也可以在桩土间加新的非线性单元来模拟真实情况。

```
316 equalDOF  1   97  1 2
317 equalDOF 11   78  1 2
318 equalDOF 10   59  1 2
319 equalDOF  4  101  1 2
320 equalDOF 13   82  1 2
321 equalDOF 12   63  1 2
322 equalDOF  7  105  1 2
323 equalDOF 15   86  1 2
324 equalDOF 14   67  1 2
```

下面规定输出记录文件:

```
325 foreach theNode { 6 5 4 13 12 99 80 61 42 23} {
326 recorder Node -file output/node$theNode.out -time -node $theNode
    -dof 1 2 disp
327 }
328 recorder Element -ele 23 -time -file output/stress23.out -time
    material 2 stress
329 recorder Element -ele 41 -time -file output/stress41.out -time
    material 2 stress
330 recorder Element -ele 59 -time -file output/stress59.out -time
    material 2 stress
331 recorder Element -ele 77 -time -file output/stress77.out -time
    material 2 stress
332 recorder Element -ele 37 -time -file output/stress37.out -time
    material 2 stress
333 recorder Element -ele 37 -time -file output/strain37.out -time
    material 2 strain
```

以上为输出记录。下面进行重力分析:

```
334 constraints Transformation
335 numberer RCM
336 test NormDispIncr 1.E-6 25 2
337 integrator LoadControl 1 1 1 1
338 algorithm Newton
```

1.5 土-结构相互作用体系

```
339 system BandGeneral
340 analysis Static
341 analyze 3
342 puts "soil gravity nonlinear analysis completed ..."
```

以上代码为静力分析. 接着进行动力分析, 和前面算例相同, 此处没有使用 "loadConst –time 0" 命令, 所以动力分析时系统时间从 3.0 秒开始.

```
343 wipeAnalysis
344 constraints Transformation
345 test NormDispIncr 1.E-6 25 2
346 algorithm Newton
347 numberer RCM
348 system BandGeneral
349 integrator Newmark 0.55 0.275625
350 analysis Transient
351 set startT [clock seconds]
352 pattern UniformExcitation 1 1 -accel "Series -factor 3 -filePath
    elcentro.txt -dt 0.01"
353 analyze 2400 0.005
354 set endT [clock seconds]
355 puts " 完成时间: [expr $endT-$startT] seconds."
```

第 351 行设置分析计算的初始时间, 第 354 行设置分析计算的结束时间, 第 355 行统计模型的计算耗时并输出.

(三) 结果分析

7 号梁单元第三个高斯点的弯矩曲率关系曲线如图 1.5.2 所示; 地基土的 37 号

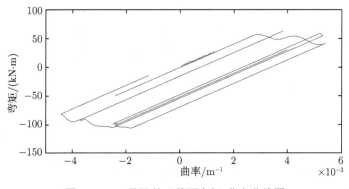

图 1.5.2 7 号梁单元截面弯矩-曲率曲线图

四节点单元的应力应变关系曲线如图 1.5.3 所示. 从图中看出, 土结体系在地震中进入强非线性阶段.

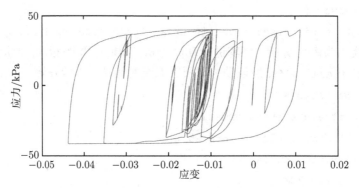

图 1.5.3   7 号地基土单元应力应变曲线图

注意   本算例中土与结构连接使用 equalDOF 命令, 即直接粘结在一起. 这是一种简化做法, 没有考虑桩的体积所占空间、桩土滑移以及土不能抗拉等性质. 在精细化模拟方法中, 必须考虑这些因素. 可采用接触单元模拟, 也可以把桩的体积用外伸的刚臂 "爪子" (梁柱单元) 模拟, 爪子外部节点与土单元节点间用法向可拉开的一维非线性弹簧和切向可滑动的非线性弹簧连接. 具体可参考 UCSD 的 Elgamal 教授的文献.

1.6   流固耦合体系

算例 1.6.1   简化的坝和库水流固耦合体系动力分析

(一) 问题简述

拱坝-库水-地基流固耦合系统非常复杂, 为了简化说明, 用图 1.6.1 的流固耦合系统. 该体系由固体单元、流体单元、固体单元和流体单元之间的接触单元、流体单元远端的透射边界单元, 和地基与大地粘弹性人工边界单元等众多单元共同组成. 实际的流固耦合系统在进行完整的动力分析需要考虑到每一种单元的作用.

注意   实际的坝和库水流固耦合体系比这个模型复杂得多, 但是基本固体和流体单元、流固耦合边界、流体边界和地基人工边界和本算例是完全相同的.

本算例主要介绍上述体系在 OpenSees 中进行动力分析的方法. 其为一个简单的块体模型, 在算例中考虑了高拱坝-库水-地基耦合体系中所有单元的共同作用. 本模型共有 16 个节点和 8 个单元, 其中单元 1(节点: 1-8) 为混凝土单元, 单元 2(节

1.6 流固耦合体系

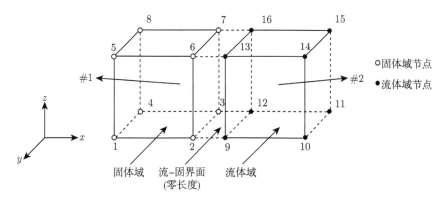

图 1.6.1 三维流固耦合系统

点: 9-16) 为库水单元. 单元 1 和单元 2 之间为接触单元, 混凝土的侧面和底部有粘弹性边界单元 (节点: 3-4-8-7, 1-2-6-5, 1-2-3-4, 1-5-8-4), 库水右端有流体透射边界单元 (节点: 10-11-15-14). 该算例将 El-centro 地震波等效为力施加在结构上, 荷载为 $P=10^6 \times$ El Centro 波的时程. 下表为块体单元中, 混凝土材料参数的取值, 混凝土密度 $\rho = 2400 \text{kg/m}^3$. 本算例 TCL 文件中单位为国际单位.

参数	大小
G	14957MPa
K	17677MPa
λ	0.11
α	26.614MPa
T	-2.0684MPa

库水密度为 $\rho = 1000 \text{kg/m}^3$, 粘弹性人工边界的参数取值为弹性模量 $E = 25$GPa, 剪切模量 $G = 10$GPa, 岩石的密度 $\rho = 2600 \text{kg/m}^3$, 法向和切向粘弹性人工边界参数 $\alpha_N = 1.33$, $\alpha_T = 0.67$.

本算例所提到的模型和计算方法可参考文献:

Y Gao, Q Gu*, Z Qiu, J Wang. 2016. Seismic Response Sensitivity Analysis of Coupled Dam-Reservoir-Foundation Systems. ASCE J of Eng Mech.

(二) 命令流分析

```
1 model basic -ndm 3 -ndf 3
```

第 1 行确定结构的维数及自由度的命令流, 其中 -ndm 3 表示三维, -ndf 3 表示每个节点有 3 个自由度.

```
2 node  1  1.00000  0.0000  0.00000
3 node  2  1.00000  1.0000  0.000
4 node  3  0.00000  1.0000  0.000000
5 node  4  0.00000  0.0000  0.00000
```

```
6 node 5 1.00000  0.0000   1.0000
7 node 6 1.00000  1.0000   1.000000
8 node 7 0.00000  1.00000  1.00000
9 node 8 0.00000  0.00000  1.0000
```

第 2-9 行定义节点的命令流。

```
10 nDMaterial TruncatedDP 1 3 2400000 1.4957e+010 1.7677e+010 0.11
   2.6614e7 -2.0684e6 1.0e-8
```

第 10 行定义材料特性的命令流。采用 TruncatedDP 材料即截断 Drucker-Prager 模型,其材料号为 1,依次定义该材料的维数、密度,以及上面表格中的参数。

```
11 set g 9.8
12 set rhog [expr -2400000*$g]
```

第 11-12 行定义体力,该材料密度为 $2400kg/m^3$。

```
13 element bbarBrick 1  1 2 3 4 5 6 7 8  1  0 0 $rhog
```

第 13 行定义块体单元的命令流。其单元号为 1,八个节点的节点标号依次为 1、2、3、4、5、6、7、8,材料号为 1,X、Y 方向体力为 0,Z 方向体力为$rhog。

```
14 recorder Element -ele 1 -time -file stress1.out -time material 1
   stress
15 recorder Element -ele 1 -time -file stress2.out -time material 2
   stress
16 recorder Element -ele 1 -time -file stress3.out -time material 3
   stress
17 recorder Element -ele 1 -time -file stress4.out -time material 4
   stress
18 recorder Element -ele 1 -time -file stress5.out -time material 5
   stress
19 recorder Element -ele 1 -time -file stress6.out -time material 6
   stress
20 recorder Element -ele 1 -time -file stress7.out -time material 7
   stress
21 recorder Element -ele 1 -time -file stress8.out -time material 8
   stress
22 recorder Element -ele 1 -time -file strain1.out -time material 1
   strain
23 recorder Element -ele 1 -time -file strain2.out -time material 2
   strain
```

```
24 recorder Element -ele 1 -time -file strain3.out -time material 3
   strain
25 recorder Element -ele 1 -time -file strain4.out -time material 4
   strain
26 recorder Element -ele 1 -time -file strain5.out -time material 5
   strain
27 recorder Element -ele 1 -time -file strain6.out -time material 6
   strain
28 recorder Element -ele 1 -time -file strain7.out -time material 7
   strain
29 recorder Element -ele 1 -time -file strain8.out -time material 8
   strain
```

第 14-29 行记录信息的命令流. 输出单元信息, 其中第 14-21 行记录单元高斯点的应力信息, 第 22-29 行记录单元高斯点的应变信息. 以第 14 行为例, 其记录的单元标号为 1, -time 表示每时步记录, 储存文件为 stress1.out, 记录高斯点 1 的应力.

```
30 set E 2.5E10
31 set G 10E10
32 set rho 2600
33 set R 475
34 set alphaN 1.33
35 set alphaT 0.67
```

第 30-35 行分别定义单元材料的信息: 弹性模量、剪切模量、质量密度、散射波到边界距离、法向粘弹性人工边界参数以及切向粘弹性人工边界参数.

```
36 element VS3D4  5 3 4 8 7  $E $G $rho $R $alphaN $alphaT
37 element VS3D4  6 1 2 6 5  $E $G $rho $R $alphaN $alphaT
38 element VS3D4  7 1 2 3 4  $E $G $rho $R $alphaN $alphaT
39 element VS3D4  8 1 5 8 4  $E $G $rho $R $alphaN $alphaT
```

第 36-39 行为定义四节点 3D 粘弹性边界单元的命令流, 其模拟底部的截断边界.

```
40 recorder Node -file node_5.out -time -precision 16 -node 5 -dof
   1 disp
41 recorder Node -file node_6.out -time -precision 16 -node 6 -dof
   1 disp
```

第 40-41 行为记录信息的命令流. 以第 40 行为例, 输出节点信息, 其储存文件为 node5.out, -time 表示每时步记录, 记录节点 5 的 X 方向的位移, 其中精确度为 16.

```
42 model basic -ndm 3 -ndf 1
```
第 42 行为建立三维、单自由度的结构.
```
43 node  9  1.00000  1.0000   0.00000
44 node 10  1.00000  2.0000   0.0000
45 node 11  0.00000  2.0000   0.000000
46 node 12  0.00000  1.0000   0.00000
47 node 13  1.00000  1.0000   1.0000
48 node 14  1.00000  2.0000   1.000000
49 node 15  0.00000  2.00000  1.00000
50 node 16  0.00000  1.00000  1.0000
```
第 43-50 行为定义节点的命令流.
```
51 set water 2
52 nDMaterial AcousticMedium $water 2.069857690E+09 1000
```
第 51-52 行为定义材料特性的命令流. 采用 AcousticMedium 材料, 其材料号为 2, 体积模量为 2.069857690E+09, 质量密度为 1000kg/m^3.
```
53 element AC3D8  3  9 10 11 12 13 14 15 16 $water
54 element ASI3D8 2  2  3  6  7  9 12 13 16
55 element AV3D4  4 10 11 15 14 $water
```
第 53-55 行分别为定义水单元、流固接触单元、水的边界单元的命令流.
```
56 fix 13 1
57 fix 14 1
58 fix 15 1
59 fix 16 1
```
第 56-59 行为定义边界条件的命令流. 以第 56 行为例, 约束节点水压力自由度.
```
60 recorder Node -file node_9.out  -time -precision 16 -node  9
   -dof 1 2 3 disp
61 recorder Node -file node_10.out -time -precision 16 -node 10
   -dof 1 disp
62 puts "Define model acoustic ok...."
63 timeSeries Path 1 -dt 0.01 -filePath elcentro.txt -factor 3
```
将 tcl 文件和地震波文件 elcentro.txt 以及 OpenSees.exe 放在同一文件夹里运行.

第 63 行为定义加载路径的命令流, 从 -filePath 文件中读取加速度矢量. 其序列号为 1, 时间步长为 0.01, 文件名为 elcentro.txt, 荷载因子为 3.
```
64 pattern Plain 111 1 {
65 load 5 1e6 0 1e6
```

1.6 流固耦合体系

```
66 load 6 1e6 0 1e6
67 load 7 1e6 0 1e6
68 load 8 1e6 0 1e6
69 }
```

第 64-69 行为施加节点荷载的命令流. 以第 65 行为例, 在节点 $5X$、Z 方向均施加 1e6N 的力. 对于实际坝体-地基系统, 外力通常作用在地基上.

```
70 wipeAnalysis
71 constraints Transformation
72 system BandGeneral
73 numberer Plain
74 test NormDispIncr 1.0E-5 20 2
75 algorithm Newton
76 integrator Newmark 0.5 0.25
77 analysis Transient
```

以上代码为动力分析.

```
78 set startT [clock seconds]
79 analyze 1600 0.01
80 puts "Dynamic analysi done..."
81 set endT [clock seconds]
82 puts "Execution time:  [expr $endT-$startT] seconds."
```

第 78 行设置分析计算的初始时间, 第 81 行设置分析计算的结束时间, 第 82 行统计模型的计算耗时并输出.

(三) 结果分析

单元 1 共有 8 个高斯点, 用应力第一不变量为横坐标, 偏应力第二不变量的二范数为纵坐标, 其应力路径如图 1.6.2 所示.

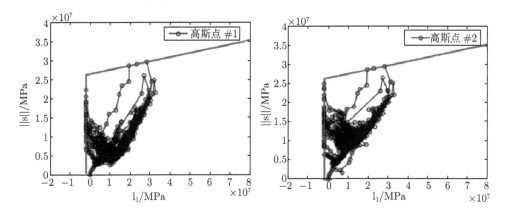

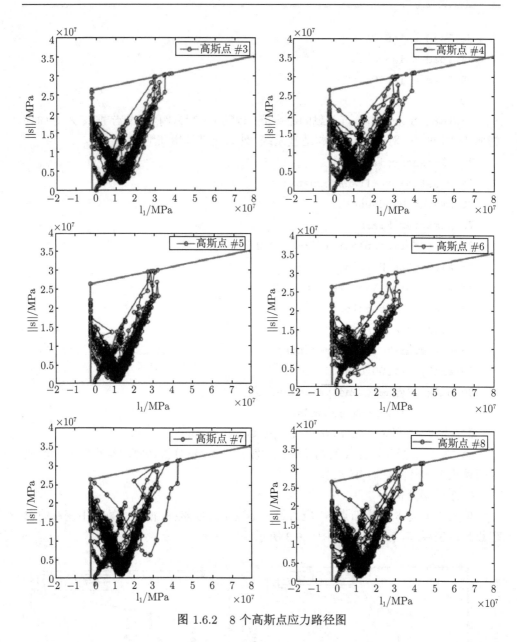

图 1.6.2 8 个高斯点应力路径图

1.7 砂土液化模型

算例 1.7.1 往复加载下饱和砂土边界面液化模型 (Bounding surfaces)

(一) 问题简述

砂土在往复荷载或地震作用下，由于孔隙水压力上升和有效应力减小，导致砂

1.7 砂土液化模型

土颗粒丧失粒间接触压力以及相互之间的摩擦力, 从而降低或失去抗剪能力, 使砂土发生液化现象. 砂土液化的危害性极大, 例如砂土液化导致建筑物以及桥梁下沉, 甚至倒塌, 或在水利工程中引起溃坝等. 砂土液化机理比较复杂, 主要表现为砂土骨架在剪切作用下发生不可逆的体积收缩 (剪缩) 和体积膨胀 (剪胀), 同时孔隙水又未能及时排除, 使得砂土失去抗剪能力, 便由固体状态转化为液体状体. 为简化说明, 采用图 1.7.1 中的往复荷载作用下砂土模型在对液化现象进行阐述.

该模型底部采用固定支座, 水位线设在单元顶部, 即直线 3-4, 以保证砂土模型充分饱和. 顶部节点 3 和 4 采用 EqualDOF, 使模型在往复荷载作用下处于纯剪状态.

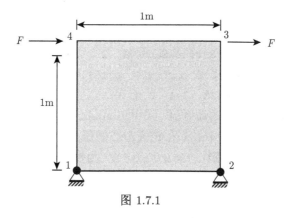

图 1.7.1

该算例加载时程见图 1.7.2. 砂土模型的材料取值分别为, 密度 $= 2\,\text{ton}/\text{m}^3$, 初始的孔隙比 $e_{in} = 0.818$, 参考剪切模量 $G_0 = 200\text{Pa}$, 泊松比 $v = 0.05$, 摩擦角 $=31°$, 模型参数 $h_r = 0.1811$, $k_r = 0.5423$, $R_f = 0.5423$, $\gamma = 0.934$, $\lambda = 0.019$, $\xi = 0.7$, $m = 3.5$, $n = 0.75$, $a = 0$, $b = 1$, $d = 1.9$.

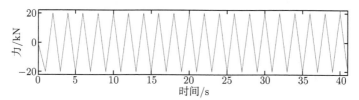

图 1.7.2 往复加载时程

本算例所提到的模型和计算方法可参考文献:

Gu Q, Wang G*. 2013. Direct Differentiation Method for Response Sensitivity Analysis of a Bounding Surface Plasticity Soil Model. Soil Dynamics and Earthquake Engineering, 49: 135–145.

Gu Q, Conte J P*, Yang Z and Elgamal A. 2009. Response Sensitivity Analysis of a Multi-Yield-Surface J2 Plasticity Model by Direct Differentiation Method.

Computer Methods in Applied Mechanics and Engineering, 198(30-32): 2272-2285.

(二) 命令流分析

```
1 model basic -ndm 2 -ndf 3
```

第 1 行为确定模型的维数及自由度的命令流,其中 -ndm 2 表示三维,-ndf 3 表示每个节点两个位移自由度和一个孔隙水自由度.

```
2 node 1 0.0 0.0
3 node 3 0.0 1.0
4 node 2 1.0 0.0
5 node 4 1.0 1.0
```

第 2-5 行为定义节点的命令流.

```
6 nDMaterial BoundingSurfaceSand 1 2 1.90 0.5066 200.0 0.818
  0.05 0.5423 0.75 0.0 1.0 1.9 0.1811 101.3250 1.01537 0.934
  0.019 0.7 3.5 0.75 0.4 0.0 0.0
```

第 6 行为定义材料特性的命令流.采用 BoundingSurfaceSand 材料,其材料号为 1.自由度为 2,密度为 1.9 ton/m³,定义最小围压 0.5066Pa,剪切模量为 200Pa,初始的孔隙比 ein=0.818,泊松比 $v = 0.05$,0.5423 为控制有效应力变化的参数,0.75 定义了破坏面与边界面大小的比值 Rp/Rf,0.0 和 1.0 为默认材料参数,1.9 为与循环加载下土的阻抗强度 (cyclic resistance) 有关的参数.模型参数 hr =0.1811,101.3250 为大气压强,其余材料参数分别为 Rf =1.01537,$\gamma = 0.934$,$\lambda = 0.019$,$\xi = 0.7$,$m = 3.5$,$n = 0.75$,$a = 0$,$b = 1$,$d = 1.9$.

```
7 element quadUP 1 1 2 4 3 1.0 1 2.2e6 1 5.09e-8 5.09e-8
  0.0 -480.0 0
```

第 7 行为定义流固充分耦合单元,其单元号为 1,四个节点的节点标号依次为 1、2、4、3,材料号为 1,X 方向体力为 0,Y 方向体力为 -480.

```
8 recorder Element -ele 1 -time -file stress.out material 1 stress
9 recorder Element -ele 1 -time -file strain.out material 1 strain
```

第 8-9 行为记录信息的命令流.输出单元信息,第 8 行记录单元高斯点的应变信息,第 9 行记录高斯点 1 的应力.

```
10 fix 1 1 1 0
11 fix 3 0 0 1
12 fix 2 1 1 0
13 fix 4 0 0 1
```

第 10-13 行定义模型的边界条件,节点 1 和 2 底部采用固定支座,节点 3 和 4 固定空隙水压力,即定义水位线.

```
14 equalDOF 3 4 1 2
```

1.7 砂土液化模型

第 14 行定义模型约束条件,使节点 3 和节点 4 在 X 和 Y 方向的位移相同,以保证模型在往复荷载作用下为纯剪应力状态.

```
15 numberer RCM
16 system ProfileSPD
17 test NormDispIncr 1.0e-8 50 1
18 algorithm KrylovNewton
19 constraints Transformation
20 integrator Newmark 1.5 1.
21 analysis VariableTransient
22 analyze 5 5.0e3 [expr 5.0e3/100] 5.0e3 20;
```

21 行是变时步加载,需要与 analyze 配合使用;

22 行是分析命令,分析 5 步,步长取很大 (5000),近似于静力加载. 每一时步如果超过 20 次迭代,就自动更改时间步长.

第 15 到 22 行为模型在自重下的静力线弹性分析.

```
23 updateMaterialStage -material 1 -stage 1
24 analyze 3 5.0e3 [expr 5.0e3/100] 5.0e3 20;
```

第 23 和第 24 行为更改材料状态从线弹性变为塑性,并执行分析.

```
25 wipeAnalysis
```

第 25 行为清除以上分析,并开始执行往复荷载.

> **注意** 第 10-14 行为砂土液化问题常用的边界条件,可扩展到大规模的计算模型中. 这里注意 OpenSees 的 UP 单元中水的自由度不是水压,而是压力的积分 (为了得到对称的刚度矩阵,可参考 Zienkiewicz 的有限元书),因此固定这一自由度相当于加 0 压力. 而指定其他形式的压力边界比较困难,必须指定一个 Series, 其值为压力积分,然后使用 pattern 命令施加外力 (比如 multi-support). 第 20-22 行用动力分析近似重力 (静力) 分析,其中材料为线弹性. 重力分析结束后,用 23 行的 updateMaterialStage 命令后,材料才变为非线性,继续动力分析 (即 24 行分析步),此时系统时间为 4000s. 动力分析也可以用基底一致激励输入方式施加地震荷载.

```
26 set P_max 20.0
27 pattern Plain 1 "Series -time {40000.0 40001.0 40002.0 40003.0
   40004.0 40005.0 40006.0 40007.0 40008.0 40009.0 40010.0 40011.
   40012. 40013. 40014. 40015. 40016. 40017. 40018. 40019.
   40020. 40021. 40022. 40023. 40024. 40025. 40026. 40027.
   40028. 40029. 40030. 40031. 40032. 40033. 40034. 40035.
```

40036. 40037. 40038. 40039. 40040. 40041.} -values {0.0
-1.0 1.0 -1.0 1.0 -1.0 1.0 -1.0 1.0 -1.0 1.0 -1.0 1.0
-1.0 1.0 -1.0 1.0 -1.0 1.0 -1.0 1.0 -1.0 1.0 -1.0 1.0
-1.0 1.0 -1.0 1.0 -1.0 1.0 -1.0 1.0 -1.0 1.0 -1.0}" {
load 3 $P_max 0.0 0
load 4 $P_max 0.0 0}

第 26 和 27 行为定义加载路径的命令流, 具体加载路径为图 1.7.2. 在节点 3 和节点 4 的 X 方向均施加 20kN 的力.

28 constraints Transformation

29 test NormDispIncr 1.0e-8 50 0

30 numberer RCM

31 algorithm Newton

32 system BandGeneral

33 rayleigh 0.0 0.0 0.02 0.0

34 integrator Newmark 0.6 [expr pow(0.6+0.5, 2)/4]

35 analysis Transient

36 analyze 4000 0.01

第 28 到第 36 行对该模型执行准静态往复加载分析.

(三) 结果分析

单元 1 共有 4 个高斯点, 以其中高斯点 1 的应力状态为例, 见图 1.7.3 和图 1.7.4.

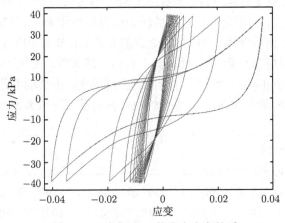

图 1.7.3 高斯点 1 的应力应变关系

1.7 砂土液化模型

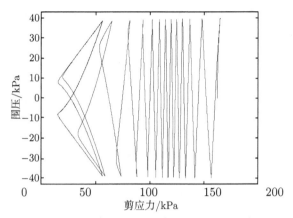

图 1.7.4 高斯点 1 的围压和剪应力关系

算例 1.7.2 往复加载下饱和砂土多屈服面液化模型 (PressureDependMultiYield)

(一) 问题简述

该算例采用与算例 1.7.1 相同的模型,只改变其材料为与围压无关的多屈服面砂土液化模型. 模型的参数为密度 $=2\text{ton/m}^3$,参考剪切模量为 60MPa,体积模量 240Mpa,摩擦角为 31 度.

(二) 命令流分析

```
1 nDMaterial PressureDependMultiYield 1 2 2 6.e4 2.4e5  31, 0.1 80
   0.5 26.5 0.1 0.2 5 10 0.015 1.
```

第 1 行为定义材料特性的命令流. 采用 PressureDependMultiYield 材料,其材料号为 1,自由度为 2,饱和土密度为 2ton/m^3,剪切模量和体积模量分别为 6.e4 和 2.4e5,摩擦角为 31 度,最大剪应变为 0.1,0.5 表示剪切模量以及体积模量随当前围压的变化关系。参考围压为 80kPa,相变角为 26.5 度,0.1 为与剪缩和孔隙水压力相关的参数,0.2 和 5 为与剪胀相关的参数,10,0.015 和 1 为与液化相关的参数。(Elgamal, Ahmed, et al. "Modeling of cyclic mobility in saturated cohesionless soils." International Journal of Plasticity 19.6 (2003): 883–905.)

(三) 结果分析

单元 1 共有 4 个高斯点,以其中高斯点 1 的应力状态为例,见图 1.7.5 和图 1.7.6.

OpenSees 中土的模型可用于复杂的液化分析和土结相互作用分析,其中土的边界条件可以按照本算例方式添加. 通常可以用基底激励方式加地震.

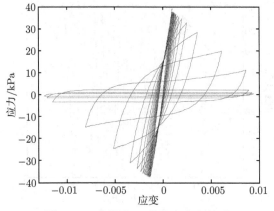

图 1.7.5　高斯点 1 的应力应变关系

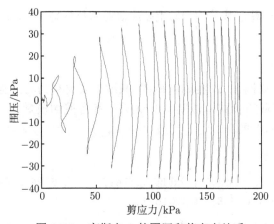

图 1.7.6　高斯点 1 的围压和剪应力关系

1.8　数值优化

在土木工程中,把数值优化方法和非线性有限元结合起来可运用于许多领域,包括结构有限元模型更新、结构识别、结构优化和可靠性度分析等.

目前 OpenSees 中有内嵌自己的优化程序:SNOPT(sparse nonlinear optimization software package),可用于各种土木工程的优化问题中. 这套 OpenSees-SNOPT 软件框架有三个突出特点:

1. 功能强大. 这是由于 OpenSees 能够很好地分析和模拟非线性结构或土-结体系,同时 SNOPT 在解决大规模非线性优化问题方面有强大的功能;

2. OpenSees 和 SNOPT 的结合非常高效. 基于 OpenSees 先进的面向对象的软件框架,SNOPT 能够很好地被封装并且嵌入 OpenSees 中,从而实现这两个系统

之间高效的数据交流和协作;

3. OpenSees-SNOPT 能广泛用于解决土木工程领域中许多优化问题. 利用 OpenSees 输入和输出的灵活性, 用户可以定义适合自己需要的目标函数和约束函数, 这大大提高了 OpenSees-SNOPT 的灵活性和应用范围.

本算例所提到的模型和计算方法可参考文献:

Gu Q, Barbato M, Conte J P*, Gill P E, and McKenna F. 2012. OpenSees-SNOPT Framework for Finite Element-Based Optimization of Structural and Geotechnical Systems. Journal of Structural Engineering, (ASCE), 138, 6.

1.8.1 基于 SNOPT 优化

SNOPT 优化的基本流程如下:

如图 1.8.1 所示, SNOPT 分析的 tcl 命令流由六个主要步骤组成, 其中 objectiveFunction 和 constraintFunction 分别指定计算目标函数和约束函数的 tcl 文件. SNOPT 优化过程中需要的结构分析、目标函数计算和约束计算等重要功能均由独立的 tcl 文件完成, 因此需要主文件和三个其他文件共四个 tcl 文件.

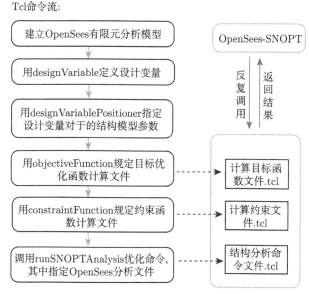

图 1.8.1 SNOPT 优化流程图

下面介绍 6 个新加入的优化命令用法.

(1) 设计变量定义的命令流为

designVariable $gradNumber -name $nameDV $startPoint-lowerBound $lBound-upperBound $upperBound

$gradNumber：设计变量标号；

$nameDV：用户指定的该变量唯一的 TCL 变量名称；

$startPoint：优化开始时的初始点（即优化变量的初始值）；

$lBound：该变量取值的下边界；

$upperBound：该变量取值的上边界。

(2) 指定设计变量位置命令流的定义为

designVariablePositioner $pos -dvNum $gradNumber -element 1 -section-Aggregator 2 -section 1 -material 1 $h

designVariablePositioner 指定了设计变量所对应的模型参数。如：

$pos：设计参数的标识号，$pos 必须从 1 开始且持续加 1；同一个设计变量可以对应多个参数。$pos 在整个模型中统一编号。

$gradNumber：设计变量的标号，由"designVariable"命令提前创建；

$ele：单元的标号，在有限元模型中由"element"命令提前创建；

$secAgg：在有限元模型中由"section Aggregation"已创建的截面组合标号；

$mat：在有限元模型中由"material"命令提前创建的材料标号；

$h：定义的设计变量，在 TCL 里为一个字符串变量。

(3) 目标函数定义的命令流为

objectiveFunction $tag -name $nameF -tclFile $tclFile -lowerBound $lBound -upperBound $uBound < -gradientName $gradF >

$tag：目标函数的标号，从 1 开始；目前 SNOPT 版本只能有一个目标函数。

$nameF：目标函数名称，为 Tcl 全局唯一的变量；

$tclFile：用户提供的一个 TCL 文件，用于计算目标函数（即：运行此 tcl 文件可以计算目标函数的值，并存在全局变量 $nameF 中），该 TCL 文件的名称为 $tclFile；另外，如果用户使用 < -gradientName $gradF >，则此文件也同时负责计算出目标函数的梯度。

$lBound：目标函数取值的下边界；

$upperBound：目标函数取值的上边界；

$gradF：存储目标函数梯度的全局变量名（可为一个数组），可选项。如果选择了此功能，则此全局梯度变量也必须在 $tclFile 计算给出。

(4) 约束函数定义的命令流为

constraintFunction $tag -name $nameG <-gradientName $gradG> -tclFile $tclFile -lowerBound $lBound -upperBound $uBound

$tag：约束函数的标号，从 1 开始；SNOPT 可以有多个约束。

$nameG：约束函数名称，为 Tcl 全局唯一的变量；

$gradG: 存储约束函数梯度的全局变量名（可为一个矩阵），可选项. 如果选择了此功能，则此全局约束梯度变量也必须在下面 $tclFile 计算给出.

$tclFile: 用户提供的一个 TCL 文件，用于计算约束函数（即：运行此 tcl 文件可以计算约束函数的值，并存在全局变量$nameG 中），该 TCL 文件的名称为$tclFile；另外，如果用户使用 <-gradientName $gradG>，则此文件也同时负责计算出约束函数的梯度，存储到$gradG 全局变量中.

$lBound: 该约束函数的下边界；

$upperBound: 该约束函数的上边界.

若约束个数为 2，则其表达方式可写成如下形式：

array set uBound{1 4.0 2 5.0}

array set lBound{1 -1e20 2 -1e20}

其表示约束函数 1 的上下边界分别是 4.0, -1e20；约束函数 2 的上下边界分别是 5.0, -1e20.

(5) 运行 SNOPT 分析的命令流为

runSNOPTAnalysis -maxNumIter $maxNumber -printOptPointX $resultFile -printFlag $printFlag < -tclFileToRun $tclFile>

$maxNumber: 优化过程中迭代的最大次数；

$resultFile: 储存优化结果的文件；

$printFlag: 输出标志，不同取值具有不同意思；

(值为 0 时，抽样分析的结果将不会输出到屏幕或者文件中

值为 1 时，每一步抽样后的结果将输出到屏幕但不保存在文件中

值为 2 时，每一步抽样后的结果将输出到屏幕，其中抽样分析再一次开始计算时所需的重要信息将储存到文件中)

$tclFile: 是运行有限元分析的 tcl 文件，典型的 tcl 文件包括：静力或动力分析，分析多少步等信息.

(6) 更新参数的命令流为

updateParameter -dv $numDV -value $newValue

$numDV: 需被更新的设计变量的标号；

$newValue: 赋予设计变量的新值；

1.8.2 实例分析

(一) 问题简述

本例主要介绍一简支梁的设计优化，如图 1.8.2 所示.

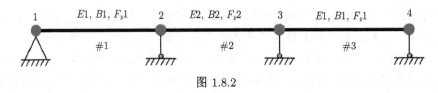

图 1.8.2

选择梁的弹性模量 ($E1$, $E2$), 刚度比 ($B1$, $B2$) 及屈服强度 (F_y1, F_y2) 作为该模型的设计变量. 将节点 4 的水平位移作为修正目标 (目标函数定义为节点 4 位移的计算值和目标值之差), 初始有限元模型通过调整设计变量得到最小化的目标函数. 各设计变量的初值和边界见下表.

	$E1$ (Pa)	$E2$ (Pa)	$B1$ (-)	$B2$ (-)	$F1$ (Pa)	$F2$ (Pa)
最大值	1e20	1e20	1e20	1e20	1e20	1e20
最小值	1e8	1e8	0.0	0.0	1e5	1e5
初值	1.8e8	1.8e8	0.016	0.016	2.7e5	2.7e5

该模型的目标函数 F 定义为

$$F = \frac{1}{2}\sum_{i=1}^{n_{\text{step}}} \left(u_i^S - u_i^R\right)^2,$$

其中 u_i 为第 i 步的节点 4 的水平位移, 上标 R 指目标时程, 上标 S 指 Opensees 计算时程, n_{step} 为时步数. 本算例中目标时程并非实测数据, 而是用另外一套设计变量取值计算得到的响应, 让 SNOPT 通过优化找到这组设计变量值.

该简支梁的设计优化包括三个 TCL 文件, 分别是主运行文件 (main.tcl)、结构分析命令文件 (tclFileToRun)、目标函数文件 (F.tcl). 因为本算例没有约束, 不用写约束函数文件. 另外本算例没有用到 designVariablePositioner 命令, 而是在每次结构分析时重新建立模型, 在模型中直接使用设计变量, 因此使文件大为简化. 除此之外, SNOPT 自带了一个 SPECS 文件 "sntoya.spc", 用于调整计算误差等, 该文本允许用户修改 (如设置新的误差限等).

(二) 命令流分析

(1) 创建 TCL 文件

TCL 文件

main.tcl　　为优化主程序

```
1 optimization
```

第 1 行: 创建优化域

```
2 designVariable 1 -name DV_E1 -startPt 1.8e8 -lowerBound 1.0E8
    -upperBound 1.e20
```

1.8 数值优化

```
3 designVariable 2 -name DV_fy1 -startPt 270000. -lowerBound
  100000.0 -upperBound 1.e20
4 designVariable 3 -name DV_b1 -startPt 0.016 -lowerBound 0.0
  -upperBound 1.e20
5 designVariable 4 -name DV_E2 -startPt 1.8e8 -lowerBound 1.0E8
  -upperBound 1.e20
6 designVariable 5 -name DV_fy2 -startPt 270000. -lowerBound
  100000.0 -upperBound 1.e20
7 designVariable 6 -name DV_b2 -startPt 0.016 -lowerBound 0.0
  -upperBound 1.e20
```

第 2-7 行: 定义设计变量的命令流. 以第 2 行为例, 其设计变量标号为 1, 变量名称为 DV_E1, 优化变量的初始值为 1.8e8, 变量的下边界为 1.0e8, 上边界为 1.e20.

```
8 objectiveFunction 1 -name F -tclFile F.tcl -lowerBound -1.e20
  -upperBound 1.e20;
```

第 8 行: 定义目标函数的命令流. 其目标函数标号为 1, 目标函数变量的名称为 F, 用户提供的 TCL 文件为 F.tcl (该 TCL 文件名恰好与目标函数变量的名称一致), F 变量的下边界为 -1e20, 上边界为 1e20.

```
9 runSNOPTAnalysis -maxNumIter 50 -printOptPointX OptX.out -print-
  Flag 1 -tclFileToRun tclFileToRun.tcl
```

第 9 行: 运行 SNOPT 分析的命令流. 其迭代的最大次数为 50, 储存优化结果文件为 OptX.out, printFlag 表示输出标志, 1 为输出, 0 为不输出, 运行的文件为 tclFileToRun.tcl.

F.tcl 计算目标函数值, 在结构分析后自动被 SNOPT 调用

```
 1 set F 0
 2 set fileId_exp [open "node4_exp.txt" "r"]
 3 set fileId_fem [open "node4.out" "r+"]
 4 while {[gets $fileId_exp  u_exp_line] >= 1} {
 5 if {[gets $fileId_fem  u_fem_line] >= 1} {
 6   set count [scan $u_exp_line "%f %e " time_exp u_exp ]
 7   if {$count != 2} {
 8     error "Error reading input - terminating script"
 9     exit;
10   }
11   set count [scan $u_fem_line "%f %e %e" time_fem u_fem u_tmp]
12   if {$count != 3} {
```

```
13      error "Error reading input - terminating script"
14      exit;
15    }
16    set F [expr $F + ($u_exp-$u_fem)*($u_exp-$u_fem)];
17  }; #if
18 }; #wihle
19 close $fileId_exp
20 close $fileId_fem
```

tclFileToRun.tcl 重新分析结构响应，此前设计变量已经更新

```
1 wipe
```

第 1 行表示删除内存里的 OpenSees 模型 (包括单元、节点)，但 TCL 变量不会删除。因此设计变量仍然在内存中保存。

```
2  model basic -ndm 2 -ndf 2
3  node 1   0.0  0.0
4  node 2  10.0  0.0  -mass 3.1741e+003 0 0.0
5  node 3  20.0  0.0  -mass 4.1741e+003 0 0.0
6  node 4  30.0  0.0  -mass 5.1741e+003 0 0.0
7  fix 1 1 1
8  fix 2 0 1
9  fix 3 0 1
10 fix 4 0 1
11 uniaxialMaterial Steel01 1 $DV_fy1 $DV_E1 $DV_b1
12 uniaxialMaterial Steel01 2 $DV_fy2 $DV_E2 $DV_b2
```

第 11-12 行: 定义材料特性的命令流。以第 11 行为例，采用单轴 Steel01 材料，其材料号为 1，屈服强度为$DV_fy1，弹性模量为$DV_E1，刚度比为$DV_b1。此三个变量为设计变量，其名字在 main.tcl 中定义。

```
13 element truss 1 1 2 0.01 1
14 element truss 2 2 3 0.02 2
15 element truss 3 3 4 0.02 1
```

第 13-15 行: 定义单元的命令流。以第 13 行为例，建立杆单元，其单元号为 1，节点 1 的节点标号为 1，节点 2 的节点标号为 2，横截面面积为 0.01，采用的材料号为 1。

```
16 timeSeries Path 1 -dt 0.02 -filePath tabas.txt -factor 9.8
```

第 16 行: 定义加载路径的命令流，从 -filePath 文件中读取加速度矢量。其序列号为 1，时间步长为 0.02，文件名为 tabas.txt，荷载因子为 9.8。

```
#---------- 定义荷载形式序列号加载放向加载路径序列号
    17 pattern UniformExcitation 1 1 -accel 1
```
第 17 行：定义荷载形式的命令流．其序列号为 1，第二个 "1" 表示加载方向为 X，第三个 "1" 表示加速度序列的号码，见第 16 行．径的序列号为 1．

```
    18 constraints Plain
    19 numberer RCM
    20 test NormDispIncr 1.e -4 25 0
    21 algorithm Newton
    22 system BandSPD
    23 integrator Newmark 0.55 0.275625
    24 analysis Transient
```
第 18-24 行：分析设置的命令流．constraints Plain 表示约束边界处理，采用一般处理．numberer RCM 表示节点自由度编号采用输入节点的顺序，为一般结构使用．test NormDispIncr 1.e -4 25 0 表示收敛准则采用位移准则，容差为 1e-4，最大迭代步为 25 步，0 表示不显示迭代过程．

algorithm Newton 表示迭代算法采用牛顿法．system BandSPD 表示矩阵带宽处理采用 SPD 处理方法．integrator Newmark 0.55 0.275625 表示采用 Newmark 算法，其中 $\beta = 0.55, \gamma = 0.275625$．analysis Transient 表示为动力分析．

```
    25 recorder Node -file node4.out -time -node 4 -dof 1 2 -precision
       16 disp
```
第 25 行：记录信息的命令流．输出节点位移信息，其储存文件为 node4.out，-time 表示每时步记录，记录节点 4 的 X、Y 方向的位移，其中精确度为 16．

```
    26 set ok [analyze 2000 0.01]
```
第 26 行：设置运行 2000 步，时长 0.01s．

```
    27 remove recorder
```
第 27 行：清除之前的 recorder．

```
    28 return $ok
```

(2) 系统缺省的 SPECS 文件

SPECS 文件指定各种选项．文件以关键词 Begin 开始并以 End 结束．具体可以参考 SNOPT 用户手册．

sntoya.spc
```
Begin Toy NLP problem
    Major Print level              000001
*                                  (JFLXBT)
    Minor print level              1
    Solution                       yes
```

Major feasibility tolerance 1.0e-6 * target nonlinear constraint violation

Major optimality tolerance 1.0e-6 * target complementarity gap

Minor feasibility tolerance 1.0e-6

System information   Yes

New superbasics   10000

Hessian   full memory

End   Toy NLP problem

(三) 结果分析

设计变量	真实值	上限	下限	初值	优化结果
E1(Pa)	2.01e8	1.e20	1.e8	1.8e8	2.01e8
fy1(Pa)	307460	1.e20	1.e5	2.7e5	3.0746e5
b1	0.02	1.e20	0.0	0.016	2.0e-2
E2(Pa)	1.05e8	1.e20	1.e8	1.8e8	1.05e8
fy2(Pa)	206460	1.e20	1.e5	2.7e5	2.0646e5
b2	0.04	1.e20	0.0	0.016	4.0e-2

优化前、优化后响应与真实响应对比如图 1.8.3 所示：

> **注意** 尽管 OpenSees-SNOPT 具有强大优化功能和灵活的用户接口，但是对于结构分析中复杂的优化问题，有时候 OpenSees-SNOPT 也无法解决。这是因为在复杂的反问题中，比如设计变量过多导致局部优化解太多、强非线性导致优化响应面过于复杂、优化域不连通等问题。这些问题需要用户基于问题的机理深入分析，而不能过分依赖优化这个数学工具。通常从简单开始，逐步复杂化是个好的方法，比如先从最敏感的 3 到 5 个设计变量开始优化，先从弱非线性开始等，需要用户多分析和总结，并最终解决问题。

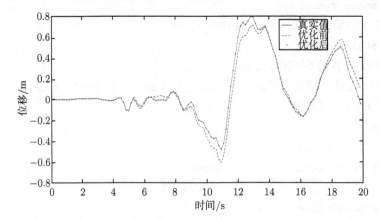

图 1.8.3　优化前、优化后响应与目标响应对比

1.9 基于 CS 技术的 OpenSees 耦合计算方法

> **注意** 在很多实际的工程问题中，有限元计算仅仅为其中一个重要环节，相当于一个"计算器"的功能，而系统中还有更上层的控制算法。这时本节的耦合和集成方法就比较重要和实用了。

有限元作为"计算器"的场合很多。比如在非线性结构优化中，优化算法将不断调整设计变量值，然后让有限元计算新的结构响应，优化算法基于这些响应来计算和更新设计变量值，这样反复迭代得到最优解。又比如，在振动台控制程序中，如果考虑上部结构试样对于振动台的影响，则必须在每一时步根据振动台的激励计算上部结构对于台面的反力，将此反力发送给控制程序，控制程序基于此信息和规定的加速度来计算下一时步应施加的控制力。

在这些过程中，有限元程序必须常驻内存，等待命令 (比如更新参数的命令和运行几步的命令)，按照命令计算完后返回结果，但并不退出内存，只是仍然继续等待 (同时把主程序控制权交给上层算法)。其功能和服务器功能类似，因此可以使用基于 Client-Server 模式的技术将 OpenSees 改造为服务器，实现这一"计算服务器"的功能。

本算例所提到的模型和计算方法可参考文献：

Gu Q, Ozcelik O*. 2011. Integrating OpenSees with other software-with application to coupling problems in civil engineering. Structural Engineering Mechanics, An International Journal, 40(1).

实例 Client-Server 模型

(一) 问题简述

本算例目标为通过 CS(Client-Server) 技术、利用 tcl 中已有的 socket 相关命令，把 OpenSees 改造为计算服务器。客户端为一段简单的 Tcl 命令，此小段命令可以集成到其他任何复杂平台中 (例如：集成到 Matlab Simulink)，一方面便于集成 (不再把整个有限元集成，只集成这一小段 Tcl 代码)，另一方面便于控制 OpenSees。算例中包括 4 个主要文件：服务器端文件 server.tcl, getTotalResistingForce.tcl, model.tcl 和客户端文件 client.tcl。集成和调用方法以及步骤如图 1.9.1 所示。

(1) 服务器端

运行下面 Tcl 代码创建服务器。首先建立结构模型，重力分析，然后驻留内存等待 (图 1.9.1)；期间如果收到 Client 传来的命令 (通过 socket 通道)，将运行此命令，(通过 socket) 返回计算结果给 Client。

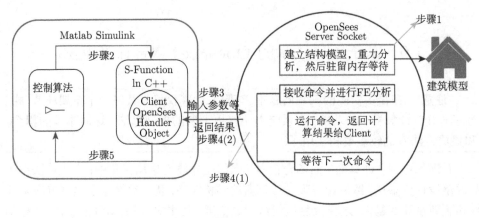

图 1.9.1

为了实现这些功能, 需要下面代码:
首先定义服务器端的接收函数 (用户简单 copy 此函数即可)

```
1 proc accept {sock ip port} {
2 fconfigure $sock -blocking 1 -buffering none
3 fileevent $sock readable [list respond $sock]
4 }
```

定义服务器的字符处理方法 (用户只需参考 10-14 行修改加入自己的内容).

```
5 proc respond {sock} {
6 if {[eof $sock] || [catch {gets $sock data}]} {
7    close $sock
8 }
```

第 6 行把从 socket 传来的客户编命令统存于 data 变量中

```
9 else {
#此处添加用户对存储在$data 中的命令和数据进行处理的方法,
10 #eval $data ; #可能的处理: 直接运行传来的命令, 比如
11 # analyze 1 0.01
11 #可能的处理: 定义全局变量 Fx, Mx, 计算反力赋给全局变量 Fx, Mx
   #并将反力通过 socket 传回客户端
12 ;#global Fx, My;
13 ;#getTotalResistingForce ;#调用用户预先定义的计算反力函数
14 ;#puts $sock "$Fx, $My"
15 return
16 }
```

1.9 基于 CS 技术的 OpenSees 耦合计算方法

```
17 }
```
如下命令将 OpenSees 改造成服务器的 tcl 主程序, 建立端口号 7200 的服务器
```
18 source model.tcl
19 socket -server accept 7200
20 vwait forever
```
(2) 客户端

作为检验, 用户可以运行另外一个 OpenSees.exe, 并逐行运行下面客户端 Tcl 代码. 更通用情况下, 这些客户端代码是由另外一个软件平台 (比如 Matlab Simulink) 调用的.

首先建立和服务器的 socket 连接, 如果服务器在本机用 localhost, 否则用服务器的 ip, 其后加端口号

```
A1.    sets [socket localhost 7200];
A2.    fconFig.$s -buffering none;
```
然后就可以向服务器发送指令了, 比如
```
A3.    puts $s "analyze 1 0.02"
A4.    gets $s
```

注意 建立 client 和 server 间的 socket 连接和通讯非常容易, 只要两台计算机之间用网络可以连接就可以实现 (或者 client 和 server 在同一台计算机上也可以). 两名计算机均需要 tcp/ip 协议.

第一, client 的 tcl 代码虽然短, 但是需要集成到其他软件平台, 或者说被其他软件调用 (比如 Matlab Simulink), 需要用其他平台的接口函数 API 实现. 通常, 将 tcl.lib 和 API 放在同一目录下就可以编译实现 API 调用 Tcl. 这里创建连接 (第 A1 A2 行) 后可以保持连接, 每次执行必要的命令 (比如第 A3、A4 行) 即可.

第二, 利用 Tcl 灵活创建函数的能力, 用户可以在服务器端创建多个函数供客户端调用. 调用的位置就在服务器端 10-14 行.

下面提供一个门式框架结构算例和编写服务器端功能函数的算例, 说明本方法的灵活性和方便性.

首先是 model.tcl 文件, 本列中建立简单门式框架结构:
```
B1 model BasicBuilder -ndm 2 -ndf 3
B2 global Me
B3 set Me [expr 7.55e6/2.0]
B4 global H
B5 global L
B6 set H 36.3
```

B7 set L 25.2

命令 B1 至命令 B7 定义了全局变量,包括了模型质量、尺寸等参数,并具体进行了赋值.

B8 node 1 0.0 0.0

B9 node 2 0.0 $H -mass $Me $Me 0.0

B10 node 4 $L 0.0

B11 node 3 $L $H -mass $Me $Me 0.0

B12 geomTransf Linear 1

命令 B8 至命令 B11 定义节点坐标,并在梁的两端节点处定义了质量,主要考虑横向与竖向两个方向的质量,不考虑弯曲的转动惯量. 命令 B12 定义了本二维模型中坐标变换,关于坐标变换的使用方法详细参见前文 1.4.2 节所述.

B13 set Area [expr 6.1*6.1]

B14 set Iz [expr 6.1*6.1*6.1*6.1/12.0]

B15 set factorK 1.0

B16 element elasticBeamColumn 1 1 2 $Area [expr 2.7e10*$factorK] $Iz 1

B17 element elasticBeamColumn 2 2 3 $Area [expr 2.0e14*$factorK] $Iz 1

B18 element elasticBeamColumn 3 3 4 $Area [expr 2.7e10*$factorK] $Iz 1

命令 B13 至命令 B15 定义与截面尺寸相关的参数,命令 B16 至命令 B18 定义两个竖向的弹性材料柱单元以及一个横向的弹性材料梁单元.

B19 constraints Transformation

B20 test NormDispIncr 1.E-6 25 1

B21 algorithm Newton

B22 numberer RCM

B23 system BandGeneral

命令 B19 至命令 B23 规定模型基本算法,以上定义模型过程中所使用的命令具体用法参见前文所述.

B24 set w1 13.9817881546;

B25 set w2 85.6248678831;

B26 set ksi 0.02

B27 set a0 [expr $ksi*2.0*$w1*$w2/($w1+$w2)]

B28 set a1 [expr $ksi*2.0/($w1+$w2)]

B29 integrator Newmark 0.5 0.25.

1.9 基于 CS 技术的 OpenSees 耦合计算方法

```
B30 rayleigh $ao  $al  0.0  0.0
B31 analysis Transient
```

命令 B24 至命令 B31 根据求得模型的前 2 阶自振周期,据此计算 rayleigh 参数. 具体在前面已讲解.

```
B32 recorder Node -file disp.out -time -node 2 3 -dof 1 2 3 disp
B33 recorder Node -file reaction.out -time -node 1 4 -dof 1 2 3
    reaction
```

命令 B32 至命令 B33 设置了输出记录文件,记录内容为顶部节点 2 与节点 3 的位移以及柱底部节点 1 与节点 4 的响应.

其次是服务端 (Server) 定义 runOpenSeesOneStep 函数的文件.

子程序是 runOpenSeesOneStep,其作用为加多点激励荷载,然后调用 OpenSees 运行一步. 具体函数定义如下:

```
C1 proc runOpenSeesOneStep {numOfTimeStep timeStep currentDisp_1
    currentDisp_2} {
```

根据函数声明格式可知本函数包括四个形参,分别代表: 加载的时步数、步长、2 个控制位移. 本例中两个位移分别代表地基中心处的水平和转动位移.

```
C2 global L
C3 remove loadPattern 1 ; # 先删除 1 号 pattern,然后在下面添加
C4 pattern MultipleSupport 1 {
```

以下 C5 至 C13 命令主要是实现对节点 1 通过位移控制进行加载.

```
C5 groundMotion 11 Plain -disp "Series -time {[expr $numOfTimeStep
    *$timeStep] } -values { $currentDisp_1}"
C6 imposedMotion 1 1 11
```

命令 C5 定义了通过多点激励 Multiple Support 进行的加载方式,使用的 Series 形式实现. Series 通常定义时间序列,此处为特殊用法,只有一个点,时间点由 $numOfTimeStep*$timeStep 确定,加载的大小由 values 后的数值确定. 之后在命令 C6 中使用 imposedMotion 命令实现,imposedMotion 命令后第一个数字 1 表示节点 1,在本算例模型中即为柱底部节点 (模型定义详见 Model.tcl). 第二个数字 1 表示第一个自由度方向,数字 11 表示之前定义的加载编号,即命令 C5 中的 11 Plain,以下命令相同.

```
C7 set currentDisp_y [expr -$currentDisp_2*$L/2 ]
C8 groundMotion 12 Plain -disp "Series -time {[expr $numOfTimeStep
    *$timeStep]} -values { $currentDisp_y } "
C9 imposedMotion 1 2 12
```

C7 到 C9 在节点 1 上施加竖向位移激励.

C10 `groundMotion 13 Plain -disp "Series -time {[expr $numOfTimeStep *$timeStep]} -values { $currentDisp_2}"`

C11 `imposedMotion 1 3 13`

C10 到 C11 在节点 1 上施加转动位移激励.

C12 `groundMotion 14 Plain -disp "Series -time {[expr $numOfTimeStep *$timeStep]} -values { $currentDisp_1}"`

C13 `imposedMotion 4 1 14`

C14 `set currentDisp_y [expr $currentDisp_2*$L/2 ]`

C15 `groundMotion 15 Plain -disp "Series -time {[expr $numOfTimeStep *$timeStep]} -values { $currentDisp_y}"`

C16 `imposedMotion 4 2 15`

C17 `groundMotion 16 Plain -disp "Series -time {[expr $numOfTimeStep *$timeStep]} -values { $currentDisp_2}"`

C18 `imposedMotion 4 3 16`

  }

至此, 基底多点激励位移已经全部添加完毕, 下面分析一步, 此步的当前系统时间恰好为多点激励所定义的时刻, 即$numOfTimeStep*$timeStep, 这样确保添加的基底激励是起作用的.

C19 `analyze 1 [expr $timeStep]`

命令 C19 表示分析一步, 计算的时间步长为$timeStep.

}

> **注意** Series -time {[expr $numOfTimeStep*$timeStep] } -values { $current-Disp_1}可能导致问题, 如前所述, 如果系统时间超出 $numOfTimeStep*$timeStep 很小 (比如 1e-16 秒), 都将导致基底位移插值变为 0 的错误产生. 因此更为稳妥的做法是定义一个时间段内的位移激励值, 比如:
>
> `Series-time{[expr$numOfTimeStep*$timeStep-1.0e-10] [expr$numOfTimeStep*$timeStep+1.0e-10]}-values {$currentDisp_1 $currentDisp_1}`

最后, 服务器端编写 **getTotalResistingForce** 函数计算支座反力. 以下对文件命令进行详细分析:

D1 `proc getTotalResistingForce {} {`

D2 `global Me`

D3 `global H`

D4 `global L`

D5 `global Fx`

```
D6  global My
D7  set Fx1 [nodeReaction 1 1]
D8  set Fx2 [nodeReaction 4 1]
D9  set Fx [expr $Fx1+$Fx2]
D10 set Fy1 [nodeReaction 2 2]
D11 set Fy2 [nodeReaction 3 2]
D12 set My [expr $Fy1*$L/2.0-$Fy2*$L/2.0-$Me*$accel_2_x*$H-$Me*
    $accel_3_x*$H]
D13 puts " new:  $Fx, $My"
    }
```

上述命令计算支座反力 (水平力和弯矩),存在全局变量 F_x 和 M_y 中.

> **注意** 当使用 Tcl 命令"nodeReaction 2 1"时,前面必须先定义 recorder 记录此节点自由度的反力,否则 nodeReaction 命令给出的是错误值.

有了这些服务器端的函数,就可以在服务器 10-14 行使用被注释的 Tcl 代码了. 其中客户端传来的命令可以是 (利用 runOpenSeesOneStep 函数)

```
A3  puts $s "runOpenSeesOneStep 1 0.01 0.015 0.002 "
A4  gets $s;#得到支座反力
```

1.10 OpenSees 的前后处理软件 GID 介绍

有限元计算平台 OpenSees 的前后处理是通过 Tcl 语言实现的,该语言是基于字符的解释型语言,但是 Tcl 命令流的建模方法会很大程度上降低建模效率,鉴于此,目前已有多种方法 (清华大学陈莉教授团队开发的 Donap、加拿大 UBC Tony Yang 教授开发的 OpenSees Navigator、陈学伟博士开发的 ETO 等) 可较方便地实现 OpenSees 的前后处理. 这些软件多是针对某类特定问题开发,对于一般的复杂模型,这些软件功能还不够强大. 本章主要介绍如何通过 GID 来实现 OpenSees 的前后处理方法. 最后,通过一个实例演示具体步骤. 注意:目前 OpenSees 官网上有一款试用软件:GID+OpenSees,用户可以试用,并结合本节介绍来学习使用.

1.10.1 GID 的基本用法

GID 是一个很好的前后处理软件,可为各个领域的数值仿真计算提供数据的导入和结果的可视化,它可以很好地处理固体和结构力学、流体动力学、电磁学、热辐射及地质力学等各个领域的问题. 目前,它被广泛的应用于高校、研究中心及企事业单位，用以发展和解决不同领域的数值仿真问题. GID 在 OpenSees 中实现前后处理的基本步骤如下：

(1) 利用 CAD 或者 GID 建模型；
(2) 通过 GID 划分网格，并导出网格数据和材料参数；
(3) 将导出的网格数据和材料参数转换成 OpenSees 能识别的 Tcl 语言；
(4) OpenSees 分析计算；
(5) 将 OpenSees 输出的计算结果转换成 GID 能识别的数据格式，然后利用 GID 进行后处理.

如上所述是用 GID 实现 OpenSees 前后处理的基本思路，在 1.10.2~1.10.3 节中将详细讲解具体实现方法. 在介绍具体操作之前，简单介绍一下 GID 的基本用法.

GID 的界面如图 1.10.1 所示，主要菜单有：1) 主菜单；2) 文件工具条；3) 命令行；4) 提示框；5) 功能工具条.

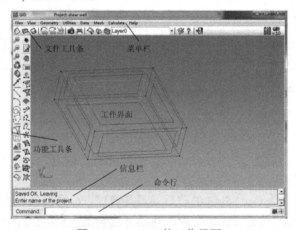

图 1.10.1 GID 的工作界面

(一) 主菜单

主菜单中：File 按钮主要包括新建、打开、导入导出文件等基本功能；View 按钮主要包含视图方面的操作功能；Geometry 按钮主要包括建模方面的基本功能；Utilities 主要包括撤销、图层、复制等基本模型操作功能；Data 包括材料参数赋值、边界条件等功能；Mesh 主要包括网格的划分方面的多种功能；Calculate 主要用于将 GID 的网格数据和材料参数传输到计算程序中，该功能在本章所讲的方法中不会用到；Help 主要包括 GID 使用方法方面的内容.

(二) 文件工具条

文件工具条中的菜单主要包括对文件的基本操作，与 File 按钮下的功能是一致的.

(三) 命令行与提示框

通过命令行,用户可以输入节点坐标等信息.

提示框是为了用户便于操作,提示错误信息、已选对象的信息等.

(四) 功能工具条

功能工具条,主要是为了用户操作方便,该工具条被 "|" 分成五块,下面将对其中的每一块分别介绍.

第一部分主要包括视图操作方面的功能,具体功能如图 1.10.2 所示. 特别要说明的是,该部分的操作只是视图操作,模型的实际位置、尺寸并不会因相关操作而有所变化. 其中,最常用的是刷新、旋转和移动.

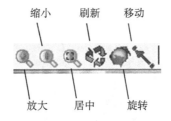

图 1.10.2 视图工具条

刚打开一个已创建的模型,或者放大 (缩小) 模型后,往往因为模型太大 (太小) 而看不到,此时通过点击刷新按钮,可以让模型调整到大小比较合适的视图.

在三维模型的前后处理过程中,可以通过旋转功能将模型调整到一个合适的角度. 具体操作为: 1) 单击旋转图标; 2) 在工作界面上单击后移动鼠标; 3) 按 "esc" 键退出该功能.

移动功能是建模过程中用到最多的按钮,它可以实现模型的移动、放大和缩小操作. 具体操作过程为: 1) 单击移动图标; 2) 在工作界面上单击; 3) 移动鼠标可对模型移动,向前 (后) 滚动滚轮可放大 (缩小) 模型; 4) 按 "esc" 键退出该功能.

第二、三部分主要包括创建线条、面和体方面的操作功能,具体地如图 1.10.3 所示. 图中 "|" 左边部分的图标可创建各种线条,右边部分的图标可创建面和体.

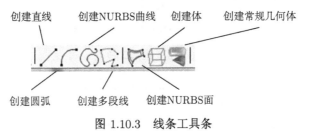

图 1.10.3 线条工具条

创建线条的基本步骤为：1) 单击对应图标；2) 根据信息栏的提示输入点的坐标或直接点击工作界面；3) 按 "esc" 键表示确认该线条的创建过程，再按 "esc" 键表示退出线条创建功能.

创建面 (体) 的操作过程与线条的相同，只是需要注意的是，创建面 (体) 时需要选择闭合的线 (面). "创建常规几何体" 图标，可快捷方便地创建圆柱、棱柱、圆锥等常规几何体，具体操作为：1) 点击图标；2) 根据信息栏的提示输入相应的参数；3) 按 "esc" 键表示确认该几何体的创建过程，再按 "esc" 键表示退出几何体创建功能.

第四部分包括删除和属性功能，如图 1.10.4 所示.

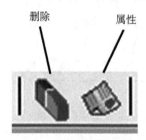

图 1.10.4　删除和属性工具条

删除功能用于删除体、面、线和点对象. 具体操作过程为：1) 点击 "删除" 图标，出现图 1.10.5 所示的删除工具条，选中相应的图标；2) 选中要删除的对象，按 "esc" 键表示确认删除，再按 "esc" 键表示退出删除功能.

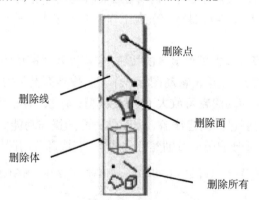

图 1.10.5　删除工具条

属性功能用于查看对象 (点、线、面和体) 的坐标、图层、边界条件等信息. 具体操作过程为：1) 点击 "属性" 图标，出现图 1.10.6 所示的属性工具条，选中相应的图标；2) 选中要查看的对象，按 "esc" 键弹出如图 1.10.7 所示的窗口，上面有所选对象的信息.

1.10 OpenSees 的前后处理软件 GID 介绍

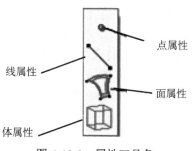

图 1.10.6　属性工具条

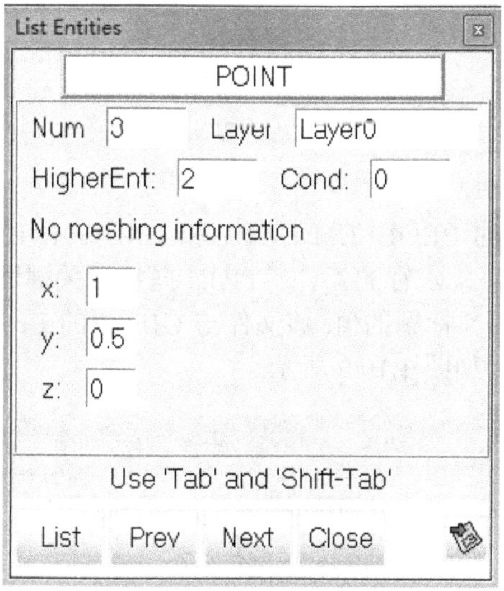

图 1.10.7　对象属性窗口

第五部分为网格操作工具条, 主要包括网格操作方面的功能, 具体地如图 1.10.8 所示. 其中, 最常用到是 "网格切换" 功能, 通过点击该图标可以实现网格和模型的切换. 该工具条中的其他图标是关于网格操作的, 对于初学者而言该部分的操作可以通过主菜单 "Mesh" 中的选项和信息栏中的提示完成.

图 1.10.8　网格操作工具条

1.10.2　OpenSees 的问题类型定义 (GID)

建立模型前, 用户需要根据所研究问题的特征在 GID 中定义问题类型 (Problem type). 这里的问题类型主要包括所研究问题的: 1) 基本参数设置和求解方法; 2) 材料参数; 3) 边界条件类型 (边界条件施加在体/面/线/点上); 4) 网格信息的输出格式. 问题类型的定义通过在 GID 安装目录中的 problemtypes 文件夹定义一个后缀名为.gid 的文件夹实现, 例如 C:\GID\problemtypes\OS.gid. 该文件夹中包括四个与文件夹同名的文件分别是 "文件夹名.prb、.mat、.cnd、.bas" (例如 OS.prb OS.mat OS.cnd OS.bas), 其中: .prb 文件用于说明基本参数设置; .mat 文件用于定义材料; .cnd 文件用于定义边界条件; .bas 用于定义网格信息的输出格式.

(一) prb 文件说明

prb 文件主要用于总体说明, 具体内容举例如下:

```
1 PROBLEM DATA
2 QUESTION: UNIT
3 VALUE: N-m-kg
4 QUESTION: Calculative method
5 VALUE: Newmark β
6 END PROBLEM DATA
```

其中: 第 1、6 行的 PROBLEM DATA 和 END PROBLEM DATA 是文件的开头和结尾关键词; 第 2、4 行的 QUESTION 是标题关键词, 其后面是标题名称; 第 3、5 行是 QUESTION 标题的值. 该文件与 GID 中的 Problem Data 窗口 (Data→Problem Data) 对应, 如图 1.10.9 所示.

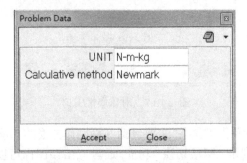

图 1.10.9　Problem Data 窗口

(二) mat 文件说明

mat 文件用于定义材料参数, 具体内容举例如下:

```
MATERIAL: uniaxialMaterial Concrete01
1 QUESTION: fc
2 VALUE: -34473.8
3 QUESTION: ec
4 VALUE: -0.005
5 QUESTION: fy
6 VALUE: -24131.66
```

```
 7 QUESTION: ey
 8 VALUE: -0.02
 9 END MATERIAL
10 MATERIAL: uniaxialMaterial Steel01
11 QUESTION: fy
12 VALUE: 248200
13 QUESTION: E
14 VALUE: 2.1e8
15 QUESTION: eab
16 VALUE: 0.02
17 END MATERIAL
```

第 1 行和第 17 行 "MATERIAL" 和 "END MATERIAL" 分别为材料的开头和结尾关键词；QUESTION 是材料的属性名称关键词, 其后面是材料的属性名称如弹性模量、泊松比等；VALUE 参数默认值关键词, 其后是材料属性的默认值. 通常, 用到的材料有多少种, 材料信息就有多少段, 如本例中有 2 种材料, 材料信息段有 2 段. 该文件与 GID 中的 Material 窗口 (Data →Material) 对应, 如图 1.10.10 所示.

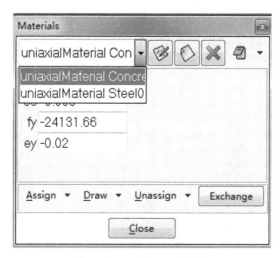

图 1.10.10 Material 窗口

在建模过程中, 单元和节点的数量很庞大而且容易出错, 而一个模型中所采用的材料往往就只有几种, OpenSees 中材料的定义完全可以通过手动的 Tcl 命令完成. 所以, 在定义材料信息段时, 不需要定义所有的材料参数名称, 也就是说每个材料信息段中只需要一个 QUESTION 行和对应的 VALUE 行. 在这里材料信息段的建立只要能说明哪一部分几何对象 (体、面、线、点) 对应哪种参数就可以.

(三) cnd 文件说明

cnd 文件用于定义边界条件，具体内容举例如下：

```
1 CONDITION: fix
2 CONDTYPE: over point
3 CONDMESHTYPE: over nodes
4 QUESTION: fixx
5 VALUE: 1
6 QUESTION: fixy
7 VALUE: 1
8 QUESTION: fixsita
9 VALUE: 1
10 HELP: fix in OpenSees manual
11 END CONDITION
```

其中：第 1、11 行是边界条件信息段的开头和结尾关键词；QUESTION 是边界条件属性关键词，其后是属性名称，如 fixx 等；VALUE 是边界条件取值关键词，其后是属性名称的默认值；CONDMESHTYPE 是边界条件类型关键词，其后为类型选项包括在体单元上 (over body elements)、面单元上 (over face elements) 以及单元节点上 (over nodes)。该文件与 GID 中的 Material 窗口 (Data →Material) 对应，如图 1.10.11 所示。

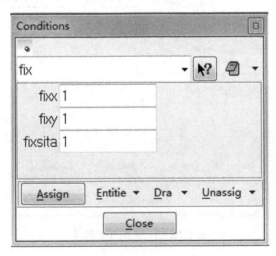

图 1.10.11 Condition 窗口

(四) bas 文件说明

bas 文件用于说明网格和材料信息的数据结构，具体内容举例如下：

```
1 # ...............node information....................
2 *loop nodes
```

1.10 OpenSees 的前后处理软件 GID 介绍

```
 3 *format "%5i%14.5e%14.5e"
 4 node *NodesNum *NodesCoord(1,real) *NodesCoord(2,real)
 5 *end nodes
 6 #....................fix informations....................
 7 *Set Cond fix *nodes
 8 *loop nodes *OnlyInCond
 9 fix *NodesNum *cond(1) *cond(2) *cond(3)
10 *end
11 #..................element informations...............
12 *set elems(all)
13 *loop elems
14 *format "%10i%10i%10i%10i%10i"
15 *ElemsNum *ElemsConec *ElemsMat
16 *end elems
```

需要特别说明的是, GID 软件系统会对 bas 文件中所有前面带 "*" 号的字符都要根据具体含义进行解析, 所有不带 "*" 号的字符, 会原封不动地输出到.dat 文件 (该文件的前处理结果文件, 可用于计算). 例如, 本例中的第 1、6 和 11 行会按照原来的格式输出到.dat 文件中. 下面详细说明各信息段的具体含义.

第 1~5 行用于说明节点信息的输出格式. 第 2 行 "*loop nodes" 表示对所有节点进行循环, 第 5 行 "*end nodes" 是对应的循环结尾标识. 第 3 行 "*format" 用于对输出的数据格式进行说明, 其后的 "%5i%14.5e%14.5e" 就是数据格式, 该格式规定规则与 C 语言的相同. 第 4 行的 node 会输出到.dat 文件中, 其后的 "*NodesNum" 用于返回节点编号, "*NodesCoord(1,real)" 和 "*NodesCoord(2,real)" 分别用于返回节点坐标的 x、y 分量.

第 6~10 行用于说明模型的边界条件. 第 7 行 "*Set Cond fix *nodes" 表示将所有边界条件名为 fix(该边界条件已在 cnd 文件定义过) 的边界条件赋到节点上. 第 8 行表示对全部有边界条件的节点循环, 第 10 行是对应的该循环的结尾标识. 第 9 行的 "fix" 会直接输出到.dat 文件中, 其后的 "*NodesNum" 表示返回节点编号, "*cond(1)、*cond(2)、*cond(3)" 分别表示返回边界条件 fix 的第 1、2、3 个属性值.

第 11~16 行用于说明单元的输出格式. 第 12 行表示后续循环中的 elems 包含所有类型的单元. 第 13 行表示对 elems 中的单元做循环, 第 16 行是该循环的结尾标识. 第 15 行的 "*ElemsNum" 用于返回单元编号, "*ElemsConec" 用于返回按照逆时针顺序排序的单元节点, "*ElemsMat" 用于返回单元的材料号.

1.10.3 OpenSees 的前处理实现方法

OpenSees 计算基于 Tcl 语言,因此在计算之前所有的材料和网格参数需按照 Tcl 语言的规则和 OpenSees 设定的关键词 (如 element、node、fix 等) 来编写. 但是,通过命令流建模型和划分网格的工作量较大,特别是建一个大规模复杂模型时. 通过 GID 建模可以较方便地得到网格信息,然后再通过 excel 或者简单的程序即可将网格信息转换成 OpenSees 能识别的格式.

(一) 建模方法

GID 建模方法有两种:1) 通过直接导入 (.dxf 或.IGES) 格式的 CAD 文件实现;2) 通过点 → 线 → 面 → 体的顺序实现,该方法将在第 1.10.4 节的实例中详细介绍.

.dxf 格式文件的导入方法为:1) 单击 Files 主菜单,如图 1.10.12 所示;2) 选择 Import 按钮;3) 选择 DXF 按钮;4) 将已完成的.dxf 或.IGES 文件导入 GID 即可,导入窗口如图 1.10.13 所示.

(二) 网格划分

GID 网格的划分通过主菜单 "Mesh" 及其子菜单实现,该主菜单的位置如图 1.10.14 所示. 如果要将面对象划分为三角形面单元或者将体划分为四面体单元,则只需定义尺寸,GID 会自动完成网格划分. 若要划分结构化的网格,则需先定义网格类型,然后定义划分份数,最后进行网格划分,该操作步骤的细节将在第 1.10.5 节中通过实例介绍.

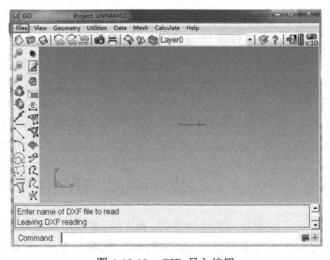

图 1.10.12 GID 导入按钮

1.10 OpenSees 的前后处理软件 GID 介绍 · 109 ·

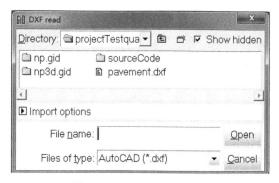

图 1.10.13　导入文件窗口

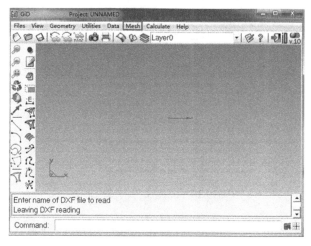

图 1.10.14　Mesh 主菜单位置

(三) 转换方法

网格划分完成后, 点击 "Calculate" 主菜单中的 Calculate 按钮, 会生成一个.msh 和.dat 文件, 这两个文件中包括模型的节点信息和单元信息, 其中.msh 是为不可读文件, .dat 是可读文件. .dat 文件中首先是节点信息, 然后是单元信息, 分别如图 1.10.15 和图 1.10.16 所示. 将该文件中的相应信息复制到 excel 中, 然后在节点号前加上 "node" 关键词, 即可完成节点部分的前处理, 在单元号前面加上 "element 单元类型名" 关键词完成单元部分的前处理.

1.10.4　OpenSees 的后处理实现方法

OpenSees 计算的计算结果为一系列文本文件, 因此还需要其他工具 (如 matlab 等) 对这些结果做一些后处理. 将这些结果写成 GID 能够识别的格式, 然后导入到 GID 可以快捷地得到云图和各种分析曲线. 相比 matlab, GID 的后处理 (尤其是云

图效果方面)功能更强大,操作更简单. 后处理的具体实现方法, 如下:

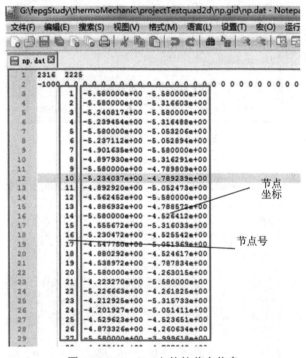

图 1.10.15 .dat 文件的节点信息

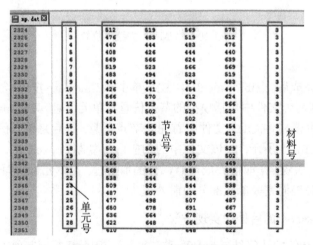

图 1.10.16 .dat 文件的单元信息

(1) 将计算结果转换为如图 1.10.17 所示的.res 文件 (为了便于显示前后关键语

1.10 OpenSees 的前后处理软件 GID 介绍

句,图中所示的文件删除了 7~2307 号节点的位移),转换过程中需特别注意文件的前后关键词.开头关键语句中 "unod0"、"Load analysis"、"Vector"、"ux"、"uy"、"uz" 为可变部分.其中:unod0 表示场(如位移、应力、温度等)的名称;Load analysis 表示对分析过程的描述;Vector 表示显示的是矢量,对张量和标量分别为 Tensor 和 Scalar;ux、uy、uz 分别表示矢量的分量值,对图中所示的文件为各个自由度方向的位移.

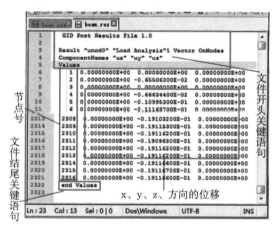

图 1.10.17 res 文件的数据格式

(2) 将.dat 文件中的数据转换成如图 1.10.18 所示的.msh 文件.特别要说明的是,.msh 文件应和.res 文件同名.

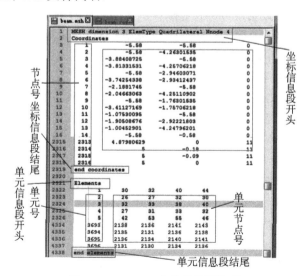

图 1.10.18 msh 文件的数据格式

(3) 点击主菜单中的前后处理切换按钮如图 1.10.19 所示，然后通过文件处理菜单 "Files" 中的 "Import"→"NASTRAN Mesh" 可导入计算结果 (.res) 和网格信息 (.msh)．可以显示云图、等值线、数据曲线等后处理结果．

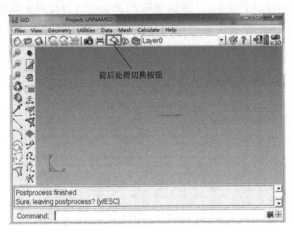

图 1.10.19　前后处理切换

1.10.5　实例

本节通过一个高、宽分别为 3m、4m 的门式框架为例，演示如何通过 GID 实现 OpenSees 的前后处理．柱梁采用非线性梁柱单元 nonlinearBeamColumn，其截面按照，弹性截面来考虑，弹性模量为 $2.49e7N/m^2$．梁的尺寸为宽 0.5m、高 1m，对应的惯性距和截面积分别为 0.04167 和 0.5．柱的尺寸为宽 0.5m、高 0.5m，对应的惯性距和截面积分别为 0.00521 和 0.25．下边界固定，左上点 A 承受水平方向的剪力，大小为 10kN．计算在该作用下的位移场．具体步骤如下：

(一) 定义问题类型文件

在 GID 中建立文件夹 beam.gid，在该文件夹内完成四个文件 (prb、mat、cnd 以及 bas)，文件的内容如下：

beam.prb 文件：

```
1 PROBLEM DATA
2 QUESTION: UNIT
3 VALUE: N-m-kg
4 QUESTION: algorithm
5 VALUE: Newton
6 END PROBLEM DATA
```

beam.mat 文件：

```
1 MATERIAL: section1
```

```
2 QUESTION: sectionNum
3 VALUE: 1
4 QUESTION: geomTransNum
5 VALUE: 1
6 QUESTION:NGasPoints
7 VALUE:5
8 END MATERIAL
9 MATERIAL: section2
10 QUESTION: sectionNum
11 VALUE: 2
12 QUESTION: geomTransNum
13 VALUE: 2
14 QUESTION: NGasPoints
15 VALUE: 5
16 END MATERIAL
```

beam.cnd 文件:

```
1 CONDITION: fix
2 CONDTYPE: over point
3 CONDMESHTYPE: over nodes
4 QUESTION: fixx
5 VALUE: 1
6 QUESTION: fixy
7 VALUE: 1
8 QUESTION: fixsita
9 VALUE: 1
10 HELP: fix in OpenSees manual
11 END CONDITION
```

beam.bas 文件:

```
1 #.................node information....................
2 *loop nodes
3 *format "%5i%14.5e%14.5e"
4 node *NodesNum *NodesCoord(1,real) *NodesCoord(2,real)
5 *end nodes
6 #..................fix informations....................
7 *Set Cond fix *nodes
```

```
 8 *loop nodes *OnlyInCond
 9 fix *NodesNum *cond(1) *cond(2) *cond(3)
10 *end
11 #...................element informations...............
12 *set elems(all)
13 *loop elems
14 *format "%10i%10i%10i%10i%10i"
15 element nonlinearBeamColumn *ElemsNum *ElemsConec *ElemsMat
*ElemsMatProp(2,int)*ElemsMatProp(3,int)
16 *end elems
```

(二) 前处理

(1) 选择问题类型"beam",具体步骤如图 1.10.20 所示.

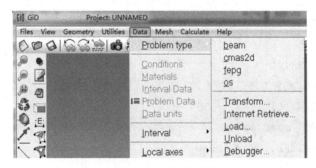

图 1.10.20　问题类型

(2) 按照点 → 线的方式建模型,建好的模型如图 1.10.21 所示.

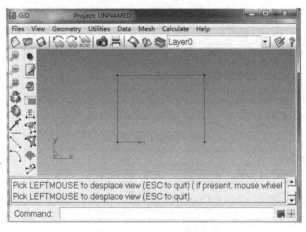

图 1.10.21　门式框架模型

(3) 将材料属性赋给几何对象. 点击 "Data→material", 选择材料, 然后按照信息栏提示, 将材料赋给线对象, 完成该操作后通过 "material→Draw→all materials" 可以查看赋值情况, 如图 1.10.22 所示.

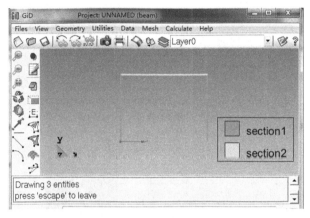

图 1.10.22 赋了材料属性后的梁柱

(4) 将边界条件赋值给固定端. 点击 "Data→conditions", 选择 fix, 然后按照信息栏提示, 将边界条件赋给点对象, 完成该操作后通过 "conditions→Draw→fix" 可以查看赋值情况, 如图 1.10.23 所示.

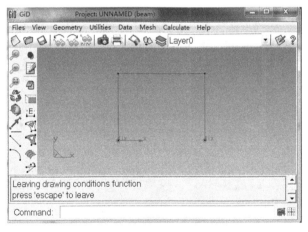

图 1.10.23 施加约束后的模型

(5) 定义网格类型和网格份数如图 1.10.24 所示, 先定义划分单元类型, 再输入划分份数 (本例中将每个梁柱分为 4 份), 然后选择要分的对象, 具体操作也可以通过查看信息栏提示完成.

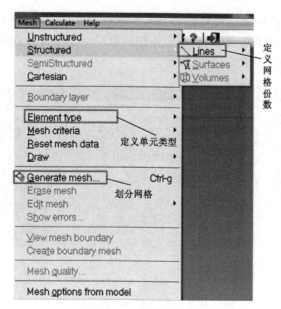

图 1.10.24 网格划分步骤

(6) 定义了网格划分情况后,可进行网格划分,具体步骤见图 1.10.24. 划分完网格后,通过点击 "View→label→All in→ 点" 可以查看网格划分情况和节点编号,如图 1.10.25 所示.

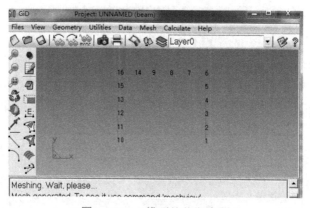

图 1.10.25 模型的节点编号

(7) 按 "ctrl+s" 将模型保存到某一目录,点击 calculate,在该目录中生成 beam.msh (二进制文件) 和 beam.dat 文件. 其中,dat 文件的内容如下:

```
# ................node information...................
1 node    1   4.00000e+00   0.00000e+00
2 node    2   4.00000e+00   6.00000e-01
```

```
 3 node    3  4.00000e+00  1.20000e+00
 4 node    4  4.00000e+00  1.80000e+00
 5 node    5  4.00000e+00  2.40000e+00
 6 node    6  4.00000e+00  3.00000e+00
 7 node    7  3.20000e+00  3.00000e+00
 8 node    8  2.40000e+00  3.00000e+00
 9 node    9  1.60000e+00  3.00000e+00
10 node   10  0.00000e+00  0.00000e+00
11 node   11  0.00000e+00  6.00000e-01
12 node   12  0.00000e+00  1.20000e+00
13 node   13  0.00000e+00  1.80000e+00
14 node   14  8.00000e-01  3.00000e+00
15 node   15  0.00000e+00  2.40000e+00
16 node   16  0.00000e+00  3.00000e+00
#..................fix informations...................
17 fix  1 1 1
18 fix 10 1 1
#..................element informations...............
19 element nonlinearBeamColumn   1  10  11  5  1  1
20 element nonlinearBeamColumn   2  11  12  5  1  1
21 element nonlinearBeamColumn   3  12  13  5  1  1
22 element nonlinearBeamColumn   4  13  15  5  1  1
23 element nonlinearBeamColumn   5  15  16  5  1  1
24 element nonlinearBeamColumn   6  16  14  5  2  2
25 element nonlinearBeamColumn   7  14   9  5  2  2
26 element nonlinearBeamColumn   8   9   8  5  2  2
27 element nonlinearBeamColumn   9   8   7  5  2  2
28 element nonlinearBeamColumn  10   7   6  5  2  2
29 element nonlinearBeamColumn  11   6   5  5  1  1
30 element nonlinearBeamColumn  12   5   4  5  1  1
31 element nonlinearBeamColumn  13   4   3  5  1  1
32 element nonlinearBeamColumn  14   3   2  5  1  1
33 element nonlinearBeamColumn  15   2   1  5  1  1
```

(8) 如果得到的.dat 文件不符合 OpenSees 语法规则, 则需要在单元信息段的前面加上单元类型关键词, 但一般情况下是符合语法规则的. 本例中不需要对 dat

文件中的单元信息段做任何修改，只需在文件的最前面加上维数和自由度说明语句（下面第 1 行），在约束信息段的后面添加如下材料信息段（第 2~3 行）和坐标转换信息段（第 4~5 行），即完成了 OpenSees 的模型部分。

```
1 model basic -ndm 2 -ndf 3
2 section Elastic  1  2.49e7  0.25  0.00521;
3 section Elastic  2  2.49e7  0.50  0.04167;
4 geomTransf Linear 1;
5 geomTransf Linear 2;
```

（三）后处理

(1) 在模型信息段的后面加上 OpenSees 加载和分析信息段（如下所示）。

```
1 recorder Node -file output/disp.out -nodeRange 1 16 -dof 1 2 3 disp;
2 pattern Plain 2 Linear {
3     load 16 10000 0.0 0.0; }
4 constraints Plain;
5 numberer Plain;
6 system BandGeneral;
7 test NormDispIncr 1.0e-8 6 2;
8 algorithm Newton;
9 integrator LoadControl 0.1;
10 analysis Static
11 analyze 10;
12 puts "pushover finished!"
```

(2) 用 OpenSees.exe 求解，并输出计算结果，然后将计算结果按照第 1.10.3 节所述内容转换成 GID 格式的后处理结果，主要是指得到 .msh 文件（可通过 files→export 得到）和与之同名的 .res 文件（通过对 OpenSees 计算结果做适当转换得到）。

(3) 将 GID 切换到后处理模式，然后读入计算结果，得到如图 1.10.26~图 1.10.28 所示位移场计算结果。

GID 功能复杂、强大，这里只讲了很简单的算例用法，作为抛砖引玉，引领读者入门之用。对于更为复杂的用法，用户需要自己查阅 GID 和 OpenSees 资料，基于本节的知识，去研究新的功能和用法，一定会事半功倍。

1.10 OpenSees 的前后处理软件 GID 介绍

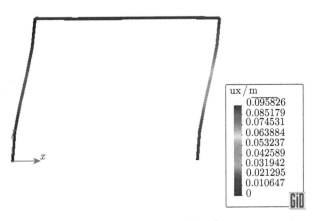

图 1.10.26　水平向位移

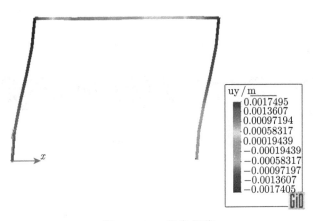

图 1.10.27　竖向位移

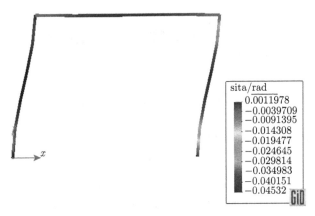

图 1.10.28　转角位移

第二部分　OpenSees 编程基础

2.1　下载与编译

2.1.1　下载 OpenSees 源代码

OpenSees 的源代码, 需要通过软件 TortoiseSVN 进行下载.

(一) 下载并安装 TortoiseSVN

进入 TortoiseSVN 的官方网站 https://tortoisesvn.net/, 如图 2.1.1 所示, 单击【Downloads】(下载), 单击下载符合当前计算机操作系统类型的 TortoiseSVN(32 位或者 64 位) 安装包, 下载完成后进行安装操作, 结束后重新启动计算机.

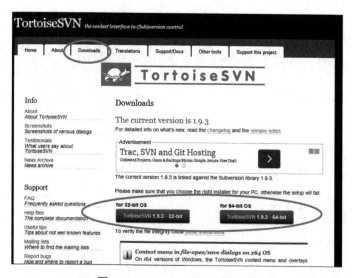

图 2.1.1　TortoiseSVN 的下载

(二) 使用 TortoiseSVN 下载 OpenSees 源代码

(1) 进入 OpenSees 的官方网站 http://opensees.berkeley.edu/, 如图 2.1.2 所示, 单击【DEVELOPER】(开发者), 然后单击【Download】(下载), 复制 svn co 后面的库 svn://peera.berkeley.edu/usr/local/svn/OpenSees/trunk OpenSees;

(2) 在桌面上新建空白文件夹 OpenSees, 鼠标右键单击该文件夹, 选择【Tortoise SVN】, 然后单击【Export】, 如图 2.1.3 所示;

2.1 下载与编译

(3) 在打开的"Export"窗口，如图 2.1.4 所示，在【URL of repository】下属空白处粘贴上面复制的库"svn://peera.berkeley.edu/usr/local/svn/OpenSees/trunk OpenSees"，如果需要下载最新版本的 OpenSees 源代码，则在【Revision】中选择【HEAD revision】；如果需要下载特定版本的 OpenSees 源代码，则在【Revision】中选择【Revision】，并在后面的空白处填入其对应的发布标签 (例如，需要下载版本为 2.4.6 的源代码，其对应的发布标签为 6123，则在空白处填 6123)，然后单击【OK】，就开始下载 OpenSees 源代码.

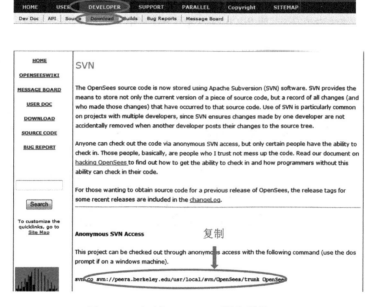

图 2.1.2 复制 OpenSees 源代码的 svn

图 2.1.3 选择 SVN 的 Export

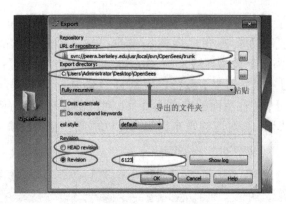

图 2.1.4 载 OpenSees 源代码

2.1.2 下载并安装 TCL

进入 OpenSees 的官方网站 http://opensees.berkeley.edu/，单击【USER】(使用者)，然后单击【Download】(下载)，在【Registered User】中输入邮箱号码进行注册，输入完成后单击【Submit】，然后页面会跳转到下载界面，单击下载符合当前计算机操作系统类型的 TCL(32 位或者 64 位) 安装包，如图 2.1.5 所示. 将下载的安装包放入计算机 C 盘目录下，双击进行安装，安装目录使用默认目录.

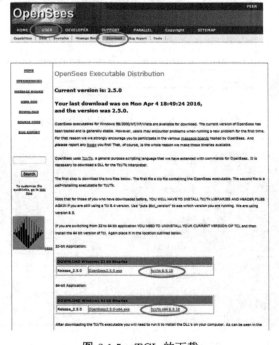

图 2.1.5 TCL 的下载

2.1 下载与编译

注意 本章编译 OpenSees 时基于 Windows7、Visual C++ 2010 的 32 或者 64 位操作系统的. 其他系统编译方法会略有区别, 但是 OpenSees 中添加的源代码是相同的.

2.1.3 下载并安装 Visual Studio 2010

进入微软官方网站 https://www.microsoft.com/zh-cn, 在右上方的搜索中搜索 Visual Studio 2010, 在搜索的结果页面中选择【Downloads】, 在【下载】中选择【Microsoft Visual Studio 2010 Service Pack 1(exe)】, 如图 2.1.6 所示. 然后选择语言并单击【下载】, 如图 2.1.7 所示.

图 2.1.6 Visual Studio 2010 的下载 (1)

此 Web 安装程序将下载并安装 Visual Studio 2010 Service Pack 1。安装过程中需要连接 Internet。有关替代 (ISO) 下载选项, 请参见下面的 "其他信息" 部分。请注意: 此安装程序适用于 Visual Studio 2010 的所有版本 (学习版、专业版、高级专业版、旗舰版、专业测试工具版)。

图 2.1.7 Visual Studio 2010 的下载 (2)

2.1.4 测试 Visual Studio 是否安装成功

安装完成后, 要先测试 Visual Studio 安装是否成功, 这里使用最简单的 "Hello world" 程序进行测试.

(一) 建立新的项目

打开 Visual Studio, 选择菜单栏中的【File】(文件)→【New】(新建)→【Project】(项目), 如图 2.1.8 所示; 在打开的 "New Project" 窗口中, 选择【Visual C++】→【Win 32 Console Application】, 然后在【Name】中输入 "test1", 单击【OK】新建项目, 如图 2.1.9 所示; 在 "test1.cpp" 中输入下述代码:

```
#include <iostream>
int main()
{
    std::cout<<"Hello world"<<std::endl;
    return 0;
}
```

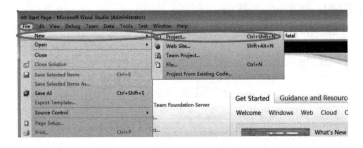

图 2.1.8 Visual Studio 新建项目 (1)

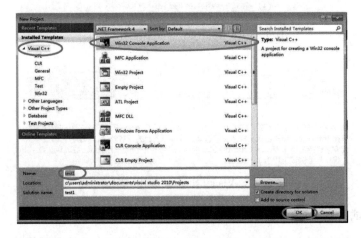

图 2.1.9 Visual Studio 新建项目 (2)

(二) 编译

单击菜单栏中的【Build】(解决方案)→【Build Solution】(生成解决方案), 如图 2.1.10 所示, 检查 "Output"(输出) 窗口中的显示是否报错, 如果报错, 说明 Visual Studio 2010 并未安装成功。

2.1 下载与编译 · 125 ·

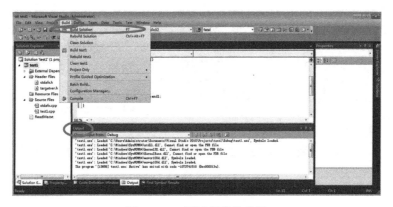

图 2.1.10 测试程序的编译

(三) 运行代码

使用 F9 在 "return 0" 所在行设置断点, 然后使用 F5 运行代码, 检查 DOS 窗口的输出是否为 "Hello world", 如图 2.1.11 所示.

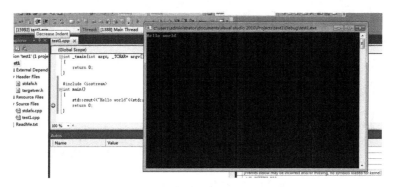

图 2.1.11 测试程序的运行

2.1.5 编译 OpenSees 源代码

编译 OpenSees 源代码, 需要打开 OpenSees 的项目文件, 添加 TCL 库, 才能进行编译.

(一) 打开 OpenSees 项目

打开 Visual Studio 2010, 单击菜单栏中的【File】(文件)→【Open】(打开)→【Project/Solution】(项目/解决方案), 如图 2.1.12 所示; 在打开的 "Open Project" 窗口中, 进入 2.1.1 节下载的 OpenSees 源代码所在的文件夹, 根据当前计算机操作系统类型选择进入 "Win32" 或者 "Win64" 文件夹, 选中其中的 "OpenSees.sln", 单击【打开】, 如图 2.1.13 所示.

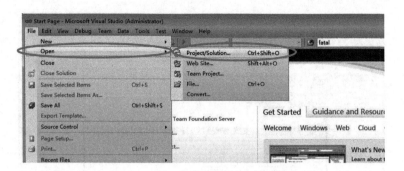

图 2.1.12　打开 OpenSees 项目 (1)

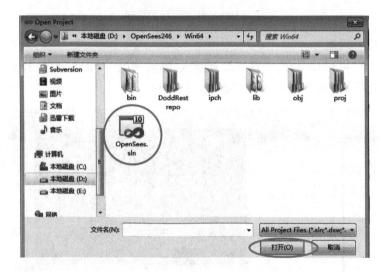

图 2.1.13　打开 OpenSees 项目 (2)

(二) 添加 TCL 库

打开菜单栏中的【View】(视图)→【Property Manager】(属性管理器), 如图 2.1.14 所示, 在打开的 "Property Manager" 窗口中选择【OpenSees】→【Debug | x64】的下拉菜单, 双击【Microsoft.Cpp.x64.user】, 如图 2.1.15 所示; 在打开的 "Microsoft.Cpp.x64.user Property Pages" 窗口中选择【Common Properties】→【VC++ Directories】, 在 "General" 中分别选择【Include Directories】和【Library Directories】下拉菜单中的【edit】, 如图 2.1.15 所示; 在打开的【Include Directories】和【Library Directories】中添加点击, 然后单击选择 TCL 安装目录下的 "include" 和 "lib" 文件夹, 单击【OK】以添加 TCL 的 "include" 和 "lib", 如图 2.1.16 和图 2.1.17 所示.

2.1 下载与编译

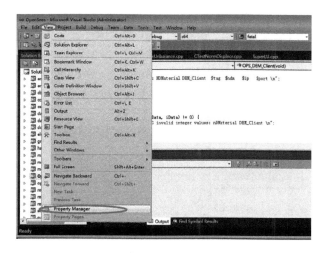

图 2.1.14　打开属性管理器

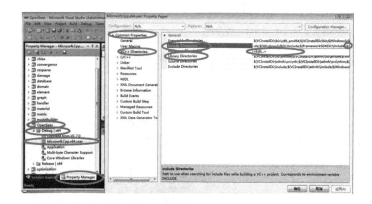

图 2.1.15　设置 VC++ Directories

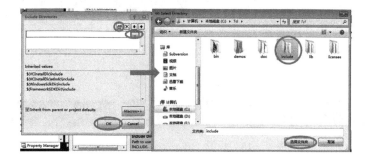

图 2.1.16　添加 TCL 的 include

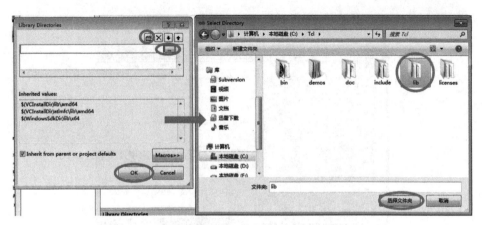

图 2.1.17　添加 TCL 的 lib

(三) 设置 OpenSees 为启动项目

打开 "Solution Explorer" 窗口，右键单击【OpenSees】，在弹出菜单中选择【Set as StartUp Project】，如图 2.1.18 所示。

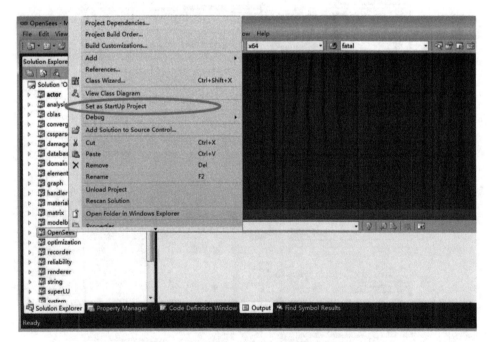

图 2.1.18　设置 OpenSees 为启动项目

(四) 编译 OpenSees

单击菜单栏中的【Build】→【Build Solution】，检查 "Output" 窗口中的显示报

错,通过上网查找错误的解决方法,修改错误后,单击菜单栏中的【Build】→【Rebuild Solution】,直到"Output"窗口显示"Build: 25 succeeded, 0 failed, 0 up-to-date, 0 skipped"为止,如图 2.1.19 所示.

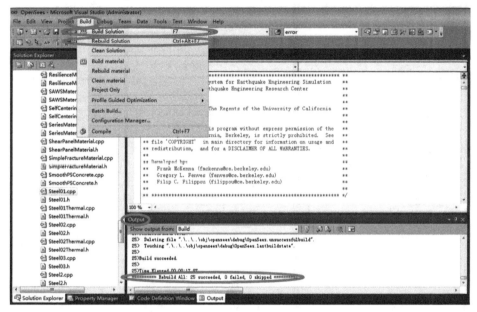

图 2.1.19   修改 error 并 Rebuild Solution

(五) 编译中的常见错误

解决编译错误的方法是,通过查找功能查找"Output"窗口的关键字"error",优先解决最前面的错误. 由于 Visual Studio 和 OpenSees 源代码版本的不同,以及计算机的差异,均会导致编译过程中报错. 故本书为简便起见,仅就使用 Visual Studio 2010 编译 OpenSees2.4.6 时经常出现的错误提供解决方案.

a. 错误 "fatal error C1083: Cannot open include file: 'elementAPI.h': No such file or directory."

这个错误的解决方法是双击"Output"窗口的这条报错记录,打开包含这条错误指令的 *.cpp 文件,将其中的 "#include< elementAPI.h>" 改为#include<D:\OpenSees246\SRC\api\elementAPI.h>,其中的 "D:\OpenSeees" 应改为本机的源代码目录.

b. 错误 "fatal error C1083: Cannot open source file: '..\..\..\SRC\material\uniaxial\Stage56Material.cpp': No such file or directory"

这个错误的解决方法是打开【Solution Explorer】→【material】→【uniaxial】,删除其中的 "Stage56Material.h" 和 "Stage56Material.cpp" 文件,如图 2.1.20 所示.

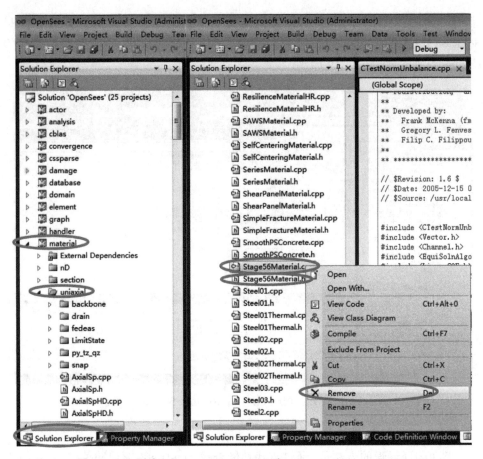

图 2.1.20　删除 "Stage56Material.h" 和 "Stage56Material.cpp"

> **注意**　不同的计算机和不同版本的 OpenSees 在编译时还可能遇到其他不同类型的错误, 这些错误很可能是由于当前版本 OpenSees 还不够完善造成的 (比如上例中新加入的 Stage56Material 还存在错误), 不要试图去调试代码, 而是尽可能回避错误. OpenSees 是个庞大的资源库, 我们只需要去研究自己所需要的那一小部分就可以了.

2.2　C++ 基本语法

　　本节概述 C++ 基本语法, 作为 OpenSees 编程基本训练之一. 建议读者把本节算例逐一输入 VC++, 编译、调试并运行, 成功后把算例保存起来以备后面查用. 学习 C++ 最好的方法只有一条: 就是实践! 注意: 对于 C++ 已经比较熟悉的用户可以跳过此节!

2.2.1 OOP 与 C++

- OOP 简介

OOP(object oriented programming) 是指面向对象的程序设计,它将现实世界中的实体归属为某类事物,这些实体是某类事物的实例,即对象,对象之间通过消息相互作用. 传统的面向过程的编程用公式可表示为 "程序 = 算法 + 数据结构". 面向对象的编程可用公式可表示为 "程序 = 对象 + 事件". 因此, OOP 的本质是模型化对象而不是数据.

对象具有属性和行为,属性是指对象的特征,例如可以用一个对象来表示哺乳动物,所有哺乳动物都具有一些属性,如年龄、颜色、敏捷程度等,同时,哺乳动物也有一些行为,如奔跑、嚎叫、进食等动作. 哺乳动物对象将描述共同属性的数据和作用于数据的操作 (体现事物的行为) 组织在一起模拟哺乳动物.

- 面向对象主要特征: 封装性、继承性、多态性

封装性: 封装使数据和加工该数据的方法封装为一个整体, 以实现模块的独立性. 用户只需知道如何使用对象而不用知道该对象是如何工作的, 从而实现数据隐藏.

继承性: 在原有类的基础上, 通过声明一个新的类, 即扩展原有的类, 实现代码的重用.

多态性: 同一消息为不同的对象接受时可产生完全不同的行动, 这种现象称为多态性. 即同一个名字所指的行为具有不同的表现形式.

- 如何学习 C++

C++ 和 OpenSees 是一个庞大的体系,系统学习并非易事. 传统的系统学习 C++ 方法一般耗时多,周期长. 建议只学习 OpenSees 二次开发所需要的部分 C++ 知识,开发者将主要精力用于算法的实现,这将事半功倍.

2.2.2 C++ 基本语法概述

C++ 程序由预处理头文件、数据 (常量、变量和对象)、函数 (包括 main() 函数和其他一般函数)、注释等组成. 每个 C++ 程序一般有两个文件. 一个文件用于保存程序的声明,包括数据成员和函数成员,以 ".h" 为后缀;另一个文件用于保存函数成员的实现,以 ".cpp" 为后缀.

- 基本符号说明

序号	命令符号	说明
1	#include<filename> 或者#include"filename"	#预处理标志，include 是一条预处理指令，表示将后面的文件复制到当前源代码文件中．<> 表示文件为标准头文件，"" 表示非标准头文件．
2	using namespace Name	指定名称空间，表示名称的搜索范围
3	int main()	每个程序只有一个主函数，即 main 函数，程序的执行从这里开始，返回值是 int 类型
4	cout << 或者 cin >>	控制台输出命令，cout 表示将字符打印到屏幕，cin 表示由屏幕输入，<< 插入运算符，>> 提取运算符．
5	return	返回值命令
6	//, /**/	注释符，//单行注释符，/**/多行注释符
7	endl	换行符

本节算例非常简单，建议用户逐行输入计算机、编译、调试、并运行出结果．将这些运行无误的算例保存起来为日后编写类似程序参考之用．关于如何输入程序和编译，请参考"2.1.4 测试 Visual Studio 是否安装成功"一节．

算例 2.2.1　C++ 程序基本组成演示

➢ 输入

```
#include <iostream>                          //包含头文件 iostream
using namespace std;                         //使用名称空间 std
int main(){                                  //主函数
    cout<<"Hello!"<<endl;                    //输出字符 Hello!，换行
    cout<<"This is a C++ program.";          //输出句子到控制台屏幕
    return 0;                                //返回值为 0
}
```

➢ 输出

```
> 
Hello!
This is a C++ program.
```

算例 2.2.2　定义和引用全局函数演示

➢ 输入

```
#include <iostream>
using namespace std;
//定义全局函数 function，参数列表 double Num1、double Num2，返回值为 double 类型
```

```
double function(double Num1,double Num2) {
    double Sum = 0.0;
    return Sum = Num1+Num2;
}
int main(){                                    //主函数
    cout<<"Enter the two numbers:"<<endl; //输入两个参数
    double Num1 =0.0,Num2 =0.0;                //声明输入参数类型并赋初值
    cin>>Num1;                                 //输入 Num1
    cin>>Num2;                                 //输入 Num2
    double Sum = function(Num1,Num2);          //引用全局函数 function
    cout<<Num1<<"+"<<Num2<<"="<<Sum<<endl; //输出说明字符和计算结果
    return 0;                                  //主函数返回值
}
```

➢ 输出

```
>
Enter the two numbers:
12
12
12+12=24
```

2.2.3 变量与常量

◆ 常量与变量

常量是指在程序运行过程中,其值保持不变的量. C++ 中的常量有三种: 字面常量、符号常量和枚举常量.

变量是指在程序运行过程中,其值可以改变的量, C++ 中变量是用名称标明的一块内存. 要使用变量,必须先进行声明,格式如下:

$$\text{data\_type var\_name}$$

在 C++ 编程时,最重要的 C++ 常量类型是在变量类型前使用关键字 const 声明的,表示将变量声明为常量,格式如下:

$$\text{const data\_type constant\_name}$$

以上两个格式表明,对使用一般变量前,必须限定类型,这个类型限定了变量能够存储的数据值类型和对内存的需求.

◆ 基本数据类型

主类型	分类型	修饰符	占用空间 (字节)	表示范围
整型	整数型 int	short	2	$-32768 \sim 32767$
		long	4	$-2^{31} \sim (2^{31}-1)$
		unsigned short	2	$0 \sim 65535$
		unsigned long	4	$0 \sim (2^{31}-1)$
		int(32 位)	4	$-32768 \sim 32767$
		unsigned int(32 位)	4	$0 \sim (2^{31}-1)$
实型	浮点型 float	无	4	$3.4E38 \sim 3.4E38$
	双精度型 double	long	8	$-1.7E308 \sim 1.7E308$
字符型	字符型 char	signed	1	$-128 \sim 127$
		unsigned	1	$0 \sim 255$
逻辑型	布尔型 bool	无	1	$0 \sim 1$

♦ **命名规则**

变量名字不能与 C++ 关键字相同，第一个字符必须是字母或下划线，不应包含除字母、数字和下划线以外的字符。C++ 对大小写敏感，所以大小写不同的变量名是两个不同的变量。除此之外，变量名字应当直观且可以拼读，最好采用英文单词或其组合，便于记忆和阅读。变量的名字应当使用 "名词" 或者 "形容词＋名词"。例如: 当前时间变量 int currenttime.

♦ **C++ 关键字**

asm	auto	break	bool	case	catch	char	class
const	const_cast	continue	default	delete	do	double	dynamic_cast
else	enum	except	explicit	extern	false	finally	float
for	friend	goto	if	inline	int	long	mutable
new	namespace	operator	private	protected	public	register	reinterpret_cast
return	short	signed	sizeof	static	struct	switch	static_cast
template	this	throw	true	try	type_info	typedef	typeid
typename	union	unsigned	using	virtual	void	volatile	while

算例 2.2.3　不同数据类型的占用内存情况演示

➢ 输入

```cpp
#include <iostream>
using namespace std;
int main(){
    cout<<"Size of short:"<<sizeof(short)<<endl; //sizeof() 变量长度
                                                 //          计算函数
    cout<<"Size of long:"<<sizeof(long)<<endl;
    cout<<"Size of unsigned short:"<<sizeof(unsigned short)<<endl;
```

```
cout<<"Size of unsigned long:"<<sizeof(unsigned long)<<endl;
return 0;                                    //返回值为 0
}
```

➢ 输出

```
>
Size of short:2
Size of long:4
Size of unsigned short:2
Size of unsigned long:4
```

2.2.4 表达式与运算符

C++ 运算符有几十种，如算术运算符、关系运算符、逻辑运算符、位运算符等，运算符的优先级是指哪一个运算符先计算；运算符的结合律是指优先级相同时，两个相邻的运算符先计算哪一个。

♦ C++ 中运算符的优先级和结合律

优先级	符号	解释	结合律
1	( )	括号操作符	从左至右
	[ ]	数组操作符	从左至右
	->	指向成员操作符	从左至右
	.	成员操作符	从左至右
2	!	逻辑非操作符	从右至左
	~	取反操作符	从右至左
	++	自增操作符	从右至左
	--	自减操作符	从右至左
	+	正号操作符	从右至左
	-	负号操作符	从右至左
	*	指针操作符	从右至左
3	*	乘法操作符	从左至右
	/	除法操作符	从左至右
	%	求余操作符	从左至右
4	+	加法操作符	从左至右
	-	减法操作符	从左至右
5	<<	右移操作符	从左至右
	>>	左移操作符	从左至右
6	<	小于操作符	从左至右
	<=	小于等于操作符	从左至右
	>	大于操作符	从左至右
	>=	大于等于操作符	从左至右

优先级	符号	解释	结合律
7	==	等于操作符	从左至右
	!=	不等于操作符	从左至右
8	&	按位与操作符	从左至右
9	^	按位异或操作符	从左至右
10	|	按位或操作符	从左至右
11	&&	逻辑与操作符	从左至右
12	||	逻辑或操作符	从左至右
13	?:	条件操作符	从右至左
14	=	赋值运算符	从右至左
	+=、-=、*=、/=	运算赋值操作符	从右至左
15	,	逗号运算符	从左至右

算例 2.2.4　运算赋值操作符演示

> 输入

```cpp
#include <iostream>
using namespace std;
int main(){
    cout<<"Enter a number:";
    int num = 0;
    cin>>num;
    num +=1;                                    //将输入值进行 += 运算符运算
    cout<<"After num += 1,num = "<<num<<endl;   //输出运算后的值
    num-=1;                                     // -= 运算符
    cout <<"After num -= 1,num = "<<num<<endl;
    num *=2;                                    // *= 运算符
    cout<<"After num *= 2,num = "<<num<<endl;
    num /=2;                                    // /= 运算符
    cout<<"After num /= 2,num = "<<num<<endl;
    num %=2;                                    // %= 运算符
    cout<<"After num %= 2,num = "<<num<<endl;
    return 0;
}
```

> 输出

>

```
Enter a number:4
After num += 1,num = 5
After num -= 1,num = 4
After num *= 2,num = 8
After num /= 2,num = 4
After num %= 2,num = 0
```

2.2.5 函数

函数是用于完成某一特定的任务的程序单元. 分为无参函数和有参函数. 有参函数使得主调函数和被调函数之间有数据传递. 主调函数可以将参数传递给被调函数, 被调函数中的结果也可以带回主调函数.

♦ 函数的组成部分

函数由四部分构成: 返回类型、函数名、参数列表和函数体, 格式如下:

```
data_type function_name(data_type var_1,data_type var_2,...)
    {
statement_1;
statement_2;
    ......
return var;
    }
```

♦ 重载函数

重载函数是指创建多个函数名和返回类型相同, 但参数不同的函数.

♦ 引用

引用是某一变量 (目标) 的一个别名, 对引用的操作与对变量直接操作完全一样. 引用能够访问相应变量的内存单元, 这使我们在编写函数时能够使用函数内部变量.

算例 2.2.5 函数的声明、调用和参数的引用演示

➢ 输入

```
#include <iostream>
using namespace std;
void send_message(void){//定义 send_message() 函数, 无参数输入, 无返回值
    cout<<"Please Enter a number! "<<endl;
}
```

```cpp
int print_message(double n){//定义 print_message() 函数，有参数输入，有
                            返回值
    cout<<"The number is"<<n<<endl;
    return 0;
}
// square()接受两个参数，&符号告诉编译器，将指向 result 的引用传递给函数
void square(double n,double& result){
    result = n*n;
}
int main(){
    send_message();              //调用 send_message 函数
    double num = 0;
    double squ = 0.0;            //声明引用时，必须同时对其进行初始化.
    cin>>num;
    print_message(num);          //调用 print_message 函数
    square(num,squ); //调用 square 函数，squ 是函数内部变量 result 的别名
    cout<<"The square is "<<squ<<endl; // 在 square() 外部使用 resu
                                       lt 的别名，输出函数内
    return 0;                    //部变量值
}
```

➢ 输出

>
Please Enter a number!
100
The number is 100
The square is 10000

算例 2.2.6 重载函数演示
➢ 输入

```cpp
#include <iostream>
using namespace std;
void square(double n,double& result_double){        //重载 square () 函数
    result_double = n*n;
}
```

2.2 C++ 基本语法

```
void square(int m,int& result_int) {        //重载 square () 函数
    result_int = m*m;
}
int main(){
    double num_double = 0.0;
    double result_double = 0.0;
    int num_int = 0;
    int result_int = 0;
    cout<<"Please Enter a number! "<<endl;
    cin>>num_double;
    num_int = int(num_double);
       square(num_double,result_double);   //求输入值的平方
    square(num_int,result_int);            //求输入值取整后的数的平方
    cout<<num_double<<'\0'<<"'s"<<'\0'<<"square is "<<result_double<<endl; //输出结果
    cout<<num_int<<'\0'<<"'s"<<'\0'<<"square is "<<result_int<<endl; //输出结果
    return 0;
}
```

> 输出

```
>
Please Enter a number!
6.7
6.7 's square is 44.89
6 's square is 36
```

2.2.6 控制程序流程

程序流程主要有三类: 顺序、选择和循环, 这三类流程就可以描述所有程序的流程.

♦ 顺序结构: 先执行 A 操作, 后执行 B 操作.

- 选择结构: 满足条件 C, 则执行 A, 否则执行 B.

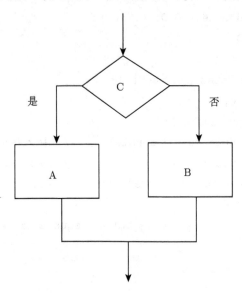

- 循环结构: 条件 C 成立时, 反复执行 A, 直到 C 为假时, 停止循环.

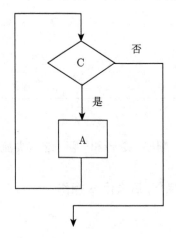

- 基本格式

2.2 C++ 基本语法

序号	命令	基本格式
1	if	if (condition)statement_1; else statement_2;
2	if_else	
3	switch	switch(expression){case pattern1: body1;break; case pattern2: body2;break; ...default:bodyN;break;}
4	while	while(condition){ statement;}
5	do...while	do{statement;} while(condition);
6	for	for (initial expression;condition ;final) {statement;}

算例 2.2.7 while 和 for 对比演示

➢ 输入

```cpp
#include <iostream>
using namespace std;
int main(){
    int i = 0;
    int sum_1 = 0;
    int sum_2 = 0;
    while (i < 6){                              //while 循环控制计算过程
        sum_1 += i;
        i++;
    }
    cout <<"1+2+3+4+5 = "<<sum_1<<endl;         //输出和值 sum_1
    for (int j = 0;j < 6;j++ ){                 //for 循环控制计算过程
        sum_2 += j;
    }
    cout<<"1+2+3+4+5 = "<<sum_2<<endl;          //输出和值 sum_2
    if((sum_1 == 15)&&(sum_2 == 15))
        cout<<"sum_1 and sum_2 are correct "<<endl;
    return 0;
}
```

➢ 输出

```
>
1+2+3+4+5 = 15
1+2+3+4+5 = 15
sum_1 and sum_2 are correct
```

算例 2.2.8　swtich 查询输入年份是否为润年并查询每月天数

➢ 输入

```cpp
#include <iostream>
using namespace std;
int main(){
    int year = 0;
    int month = 0;
    int day = 0;
    cout <<"Enter year:   ";
    cin>>year;                                          //输入查询年份
    cout<<"Enter month:   ";
    cin>>month;                                         //输入查询月份
    switch (month){
      case 1:   day = 31;break;
      case 3:   day = 31;break;
      case 4:   day = 30;break;
      case 5:   day = 31;break;
      case 6:   day = 30;break;
      case 7:   day = 31;break;
      case 8:   day = 31;break;
      case 9:   day = 30;break;
      case 10:  day = 31;break;
      case 11:  day = 30;break;
      case 12:  day = 31;break;
      case 2 :
        //if 判断是否符合润年标准
        if ((year%400 == 0)||(year%100 != 0&&year%4 == 0 )) day = 29;
        else day = 28;break;
      default:
        cout<<"The number of month is error"<<endl;break;
      }
    cout<<"The number of day is"<<day<<endl;            //输出每月天数
    return 0;
}
```

2.2 C++ 基本语法

> 输出

>
Enter year: 2000
Enter month: 2
The number of day is29

2.2.7 数组与指针

◆ 数组

数组是同一类型的一系列元素组成的完整集合, 数组元素在内存中顺序存放. 数组有两个属性, 即数组的类型和数组元素的个数. 数组在内存中的存储如下图:

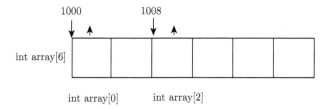

◆ 指针: 存放内存地址的变量.

定义指针是需要说明指针的类型、标志指针标识符 * 和指针名, 例如: 定义一个 data_type 指针, 并用变量 varname 的地址初始化指针, 格式如下:

$$data\_type\ *point\_name = \&varname$$

"*" 为解除引用运算符, 访问指向的数据; "&" 为引用运算符, 用来获取变量地址.

> **注意** 指针就是内存变量地址. 内存就像一个个房间, 里面空间用于存储数值, 而指针就是门牌号码. 比如 int*a=6; 等同于两句: int*a; *a=6. 第一句声明 a 是存储整数的房间的门牌号码 (内存地址), 第二句代表房间内存储的值是 6(*a=6).

算例 2.2.9 指针的使用、数组名与指针的关系演示

> 输入

```
#include <iostream>
using namespace std;
int main(){
    int Numbers[5];
```

```cpp
    int* pNumbers = &Numbers[0];        //pNumbers 指向 Numbers[0];
    int* ppNumbers = Numbers;           //ppNumbers 存储 Numbers 的值;
    cout<<"Please enter number:";       //数组元素标号是从 0 开始的;
    cin>>Numbers[0];
    cin>>Numbers[1];
    cin>>Numbers[2];
    cin>>Numbers[3];
    cin>>Numbers[4];
    cout<<"Numbers[0] is "<<*pNumbers<<endl; //输出首个元素的值
    cout<<"Numbers[1] is "<<*(pNumbers+1)<<endl; //输出第二个元素
                                                 的值
    cout<<"Numbers[2] is "<<*(pNumbers+2)<<endl; //输出第三个元素
                                                 的值
    cout<<"Numbers[3] is "<<*(pNumbers+3)<<endl; //输出第四个元素
                                                 的值
    cout<<"Numbers[4] is "<<*(pNumbers+4)<<endl; //输出第五个元素
                                                 的值
    cout<<"pNumbers is "<<'\0'<<pNumbers<<endl; //输出 pNumbers
    cout<<"*pNumbers is "<<'\0'<<*pNumbers<<endl; //输出 *pNumbers
    cout<<"Numbers is "<<'\0'<<Numbers<<endl; //输出 Numbers
    cout<<"*ppNumbers is "<<'\0'<<*ppNumbers<<endl;//输出 *ppNumbers
    cout<<"&Numbers[0] is "<<'\0'<<&Numbers[0]<<endl;//输出 *ppNumbers
    cout<<"Numbers[0] is "<<'\0'<<Numbers[0]<<endl; //输出&Numbers[0]
    return 0;
}
```

➢ 输出

>
Please enter number:1
2
3
4

```
5
Numbers[0] is 1
Numbers[1] is 2
Numbers[2] is 3
Numbers[3] is 4
Numbers[4] is 5
pNumbers is 0028FC98
*pNumbers is 1
Numbers is 0028FC98
*ppNumbers is 1
&Numbers[0] is 0028FC98 //输出结果表明数组名指向第一个元素的指针
Numbers[0] is 1
```

2.2.8 类与对象

♦ 类与对象的关系

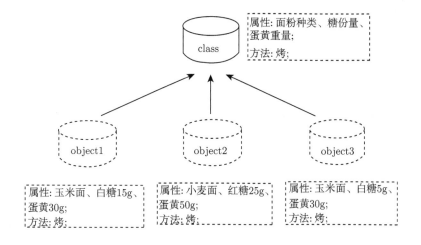

类 (class) 定义了特定的数据结构, 描述所创建对象共同的属性和方法. 当操作系统按照类的定义分配内存后, 此实体就是对象 (object). 类就像是一个做蛋糕的模子, 而对象就是所做出的蛋糕. 一个类可以生成多个对象, 具有相同的属性 (成员变量) 和方法 (成员函数). 各对象的属性和方法属于此对象本身, 在内存中互不相关. 每个对象中的成员作用域通常在对象之内 (比如: 玉米面和白糖 15g 仅限于指定对象 1 的属性, 而非对象 2), 其中 private 成员仅供内部使用, 而 public 成员可被外部调用.

♦ 类的声明

要声明类，使用关键字 class，例如声明一个猫的类：

```
class Cat{
    public:                              //关键字 public
        data_type varName_1;             //声明数据成员 varName_1
        data_type varName_2;             //声明数据成员 varName_2
        ...
        void Function_1();               //声明成员函数 Function_1()
        void Function_2();               //声明成员函数 Function_2()
        ...
};                                       //注意勿遗漏分号
```

♦ 对象的定义

类是一个类型，其成员不占用内存，只有定义了类对象后，才能对类进行操作，例如：

```
Cat Lily;                    //定义了 Cat 类的一个对象 Lliy
Lily.varName_1 = var;        //使用 ''.'' 运算符初始化数据成员 varName_1
Lily.Function_1();           //使用 ''.'' 运算符访问成员函数 Function_1()
```

♦ 类的构造函数和析构函数

类的构造函数是特殊的成员函数，只要创建了类对象，都要调用构造函数。它与类同名但不返回任何值。每当对象不在作用域或者通过 delete 销毁时，都将调用析构函数，释放动态分配的内存，实现内存资源的有效管理。

```
class Cat{
    public:
        Cat(data_type varName_1, data_type varName_2 );  //声明构造函数 Cat()
        ~Cat();                                          //声明析构函数 ~Cat()
}
```

算例 2.2.10 类的声明和类对象的定义演示

➢ 输入

```
#include <string>
#include <iostream>
using namespace std;
class Customer_Data {                     //声明一个 Customer_Data 类
```

```cpp
    private:                                  //私有数据成员
      string Name;
      int Age;
    public:                                   //公有成员函数
      Customer_Data(){                        //声明构造函数
        Name = "";
        Age = 0;
        cout<<"Constructed an object"<<endl;
      }
      ~Customer_Data(){                       //声明析构函数
        cout<<"Destructed an object"<<endl;
      }
      void SetName (string Customer_Name){    //声明 SetName() 成员函数
        Name = Customer_Name;                 //实现对私有数据成员 Name 值的设置
      }
      void SetAge (int Customer_Age){         //声明 SetAge() 成员函数
        Age = Customer_Age;                   //实现对私有数据成员 Age 值的设置
      }
      void Showdata(){                        //声明 Showdata() 成员函数
        cout<<"Name = "<<Name<<endl;          //输出私有数据成员 Name
        cout<<"Age = "<<Age<<endl;            //输出私有数据成员 Age
      }
    };
int main(){
  Customer_Data Liming; //定义 Customer_Data 类的一个对象 Liming
  Liming.SetName("Liming"); //通过公有成员函数设置私有数据成员 Name 的值
  Liming.SetAge(23); //通过公有成员函数设置私有数据成员 Age 的值
  Liming.Showdata(); //通过公有成员函数输出数据成员 Name、Age 的值
        //定义指向 Customer_Data 类对象的指针变量 pXiaofang
        Customer_Data *pXiaofang = new Customer_Data();
    pXiaofang->SetName("Xiaofang"); //使用中指针运算符 -> 访问成
                                    员函数 SetName()
    pXiaofang->SetAge(26); //使用中指针运算符 -> 访问成员函数 SetAge()
    pXiaofang->Showdata(); //使用中指针运算符 -> 访问成员函数 Showdata()
    return 0;
```

}

> 输出

```
>
Constructed an object
Name = Liming
Age = 23
Constructed an object
Name = Xiaofang
Age = 26
Destructed an object
Destructed an object
```

注 算例 2.2.10 可由 example_10.h 和 example_10.cpp 组成. 形式如下:

> 输入 example_10.h

```cpp
#include <string>
#include <iostream>
using namespace std;
class Customer_Data {                        //声明一个 Customer_Data 类
    private:                                 //私有数据成员
      string Name;
      int Age;
    public:                                  //公有成员函数
      Customer_Data(){                       //声明构造函数
      Name = "";
      Age = 0;
      cout<<"Constructed an object"<<endl;
    }
    ~Customer_Data(){                        //声明析构函数
        cout<<"Destructed an object"<<endl;
    }
      void SetName (string Customer_Name){   //声明 SetName() 成员函数
       Name = Customer_Name;                 //实现对私有数据成员 Name 值的设置
       }
      void SetAge (int Customer_Age){        //声明 SetAge() 成员函数
```

```
        Age = Customer_Age;        //实现对私有数据成员 Age 值的设置
    }
    void Showdata(){                //声明 Showdata() 成员函数
        cout <<"Name = "<<Name <<endl;  //输出私有数据成员 Name
        cout <<"Age = "<<Age<<endl;     //输出私有数据成员 Age
    }
};
```

➢ 输入 example_10.cpp

```
#include "example_010.h"       //头文件包含 example_010.h
int main(){
    Customer_Data Liming;       //定义 Customer_Data 类的一个对象 Liming
    Liming.SetName("Liming");   //通过公有成员函数设置私有数据成员 Name 的值
    Liming.SetAge(9);           //通过公有成员函数设置私有数据成员 Age 的值
    Liming.Showdata();          //通过公有成员函数输出数据成员 Name、Age 的值
    //定义指向 Customer_Data 类对象的指针变量 pXiaofang
    Customer_Data *pXiaofang = new Customer_Data();
    pXiaofang->SetName("Xiaofang"); //使用中指针运算符 -> 访问成
                                    员函数 SetName()
    pXiaofang->SetAge(26); //使用中指针运算符 -> 访问成员函数
    SetAge()
    pXiaofang->Showdata(); //使用中指针运算符 -> 访问成员函数
    Showdata()
    return 0;
}
```

➢ 输出

```
>
Constructed an object
Name = Liming
Age = 23
Constructed an object
Name = Xiaofang
Age = 26
Destructed an object
Destructed an object
```

2.2.9 继承

继承是面向对象编程的一个重要概念. 比如哺乳动物有共同特征 (属性: 年龄、体重, 方法: 奔跑、哺乳等), 而老虎除了具有这些特征外, 还具有"额外斑纹似'王'字"(属性)、捕猎 (方法) 等特征. 因此老虎类可以继承哺乳动物类, 同时加上自己的这些特征, 如下图 (哺乳动物的特征直接继承, 不用重新定义).

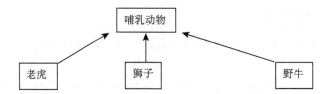

继承机制使得程序员可以创建一个类的层次结构, 每个派生类将会继承在其基类定义的方法和属性. 程序员可以对现有的代码进行快速扩展, 实现代码重用. 在 C++ 中的"继承"就是在一个已存在的类的基础上建立一个新的类. 已存在的类称为"基类"或"父类", 新建立的类称为"派生类"或"子类". 以上哺乳动物是基类, 老虎、狮子和野牛是派生类.

♦ OpenSees 中的继承的类型: 公有继承

继承方式	基类成员性质	派生类成员性质	派生类访问权限	派生类外访问权限
public	public	public	可以引用	可以引用
	protected	protected	可以引用	不可引用
	private	private	不可引用	不可引用

由此可知, 公有继承不改变基类成员在派生类中的性质且派生类外可以访问基类公有成员.

♦ 继承的基本格式

```
class ClassName :   public BaseClassName
{
  private:          //私有成员
  ...;
  public:           //公有成员
  ...;
  protected:        //保护成员
  ...;
}
```

算例 2.2.11 类的继承演示

➤ 输入

```cpp
#include <string>
#include <iostream>
using namespace std;
class Customer_Data{                        //声明一个 Customer_Data 基类
    private:                                //基类的私有数据成员
      string Name;
      int Age;
    public:                                 //基类的公有成员函数
      Customer_Data(){
        Name = "";
        Age = 0;
        cout<<"Constructed an object"<<endl;
      }
      ~Customer_Data(){
        cout<<"Destructed an object"<<endl;
    };
      void SetName (string Customer_Name) {//基类的公有成员函数 SetName()
        Name = Customer_Name;
      }
      void SetAge (int Customer_Age){      //基类的公有成员函数 SetAge ()
        Age = Customer_Age;
      }
      void Showdata(){                     //基类的公有成员函数 Showdata ()
        cout<<"Name = "<<Name<<endl;
        cout<<"Age = "<<Age<<endl;
      }
};
//声明派生类 VIP_Customer_Data 以公有继承方法继承基类 Customer_Data
class VIP_Customer_Data :    public Customer_Data{
    private:
      string Gender;                       //派生类新增的数据成员 Gender
      string Occupation;                   //派生类新增的数据成员 Occupation
```

```cpp
    public:
    void SetGender (string Customer_Gender) {
                    //新增的派生类成员函数 Gender = Customer_Gender;
        }
void SetOccpuation (string Customer_Occupation){
                //新增的派生类成员函数 Occupation = Customer_Occupation;
        }
    void Showdata(){             //派生类中与基类同名的公有成员函数
    Customer_Data::Showdata();   //使用作用域解析符::
            调用基类中成员函数 cout<<"Gender = "<<Gender <<endl;
        cout<<"Occupation = "<<Occupation <<endl;
        }
};
int main(){
    VIP_Customer_Data Liming;
    Liming.SetName("Liming"); //VIP_Customer_Data 类中没有 Setname()
    和 SetAge()
    Liming.SetAge(23);         //成员函数，但公有继承了基类的方法
    Liming.SetGender("male");  //由基类继承 Liming.SetOccpuation
                               ("worker");
    Liming.Showdata();         //派生类覆盖了基类的 Showdata() 方法
    return 0;
}
```

> 输出

```
>
Constructed an object
Name = Liming
Age = 23
Gender = male
Occupation = worker
Destructed an object
```

2.2.10 多态

多态性是面向对象的程序设计的一种重要特征,使我们可以通过调用同一个函数名,根据需要实现不同的功能. 在 C++ 中,通过继承层次结构可实现派生类的多态行为.

♦ 虚函数实现动态多态基本语法:

```
class BaseClassName
{
...
public:
    //在基类中用关键字 virtual 将成员函数声明为虚函数
    virtual data_type function (data_type var 1,...);
...
};
class ClassName :  public BaseClassName
{
...
public:
    //在派生类中具体实现该成员函数
    data_type function (data_type var_1,...){ statement;...;return value;};
...
}
```

算例 2.2.12 虚函数实现多态行为演示

➤ 输入

```
#include <string>
#include <iostream>
using namespace std;
class Pet{
    public:
        virtual void Run (){           //将基类 Pet 的 Run 方法声明为虚函数
            cout<<"Pet Run!"<<endl;
        }
};
class Dog:  public Pet{                //派生类 Dog 公有继承基类 Pet
```

```cpp
    void Run (){
        cout<<"Dog Run!"<<endl;      //派生类 Dog 改写基类中 Run () 方法
    }
};
class Cat: public Pet{               //派生类 Cat 公有继承基类 Pet
    void Run (){                     //派生类 Cat 改写基类中 Run () 方法
        cout<<"Cat Run!"<<endl;
    }
};
    //全局函数 MakePetRun() 采用引用方式调用基类 Pet 的虚函数 Run()
void MakePetRun(Pet& Name) {
    Name.Run();
}
int main(){
    Dog Bill;                        //定义 Dog 类的一个对象 Bill
    Cat Kitty;                       //定义 Cat 类的一个对象 Kitty
    MakePetRun(Bill);                //引用指向 Dog 类的对象 Bill
    MakePetRun(Kitty);               //引用指向 Dog 类的对象 Bill
    return 0;
}
```

> ➢ 输出

>
Dog Run!
Cat Run!

由程序输出结果可知，函数 MakePetRun() 没有调用基类 Pet 的方法 Run()，因为派生类 Dog 类和 Cat 类中存在改写后的虚方法 Run ()，即 Dog.Run() 和 Cat.Run()，它们优于被声明为虚函数的 Pet.Run()。因此，函数 MakePetRun() 调用的是派生类中的 Run() 方法。

2.3 OpenSees 添加新材料

了解了上述 C++ 语言的最简单和基本的语法，就可以开始学习在 OpenSees 中添加自己的程序了．这里需要进一步学习 OpenSees 程序的接口，比如添加材料、单元、算法的接口．比较好的学习方法是先阅读类似的 OpenSees 已有材料，然后

2.3 OpenSees 添加新材料

"照葫芦画瓢",就能在最快的时间内学会 OpenSees 编程. 注意: 学习 OpenSees 的方法是 "蜻蜓点水式" 的, 以实现功能为第一重要的目的, 有时有些功能函数如果不是最重要的, 可以不必逐行逐句地搞清楚 (比如 print(),setParameter() 等函数). 本节以添加新材料为例介绍 OpenSees 编程方法.

2.3.1 添加新材料背景

当现有的 OpenSees 模型不能满足科研需要时,用户可以添加自己的材料模型. 在添加新材料时选择一个材料基类作为参考, 利用 C++ 继承和多态等特属性, 即实现代码重用, 快速实现 OpenSees 的二次开发. 这里以单轴材料类 (UniaxialMaterial) 为基类, 在 OpenSees 中添加一维单轴弹性材料 (GeneralElastic). UniaxialMaterial 基类提供了派生类常用的接口. 并且该基类没有数据成员, 只有虚函数成员和一些默认的函数成员. 它们将会被派生类的同名成员函数覆盖, 即派生类在继承基类后, 将实现基类成员函数的具体行为.

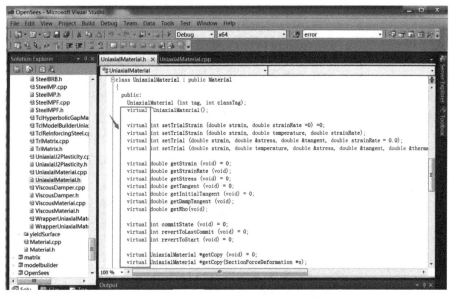

图 2.3.1 UniaxialMaterial 基类的.h 文件虚函数成员

GeneralElastic 是一维单轴弹性材料, 用户可以自定义该材料的本构包络线. 我们主要任务是当单元调用材料的成员函数时, 实现给定应变返回应力和材料弹性模量. 如图 2.3.2 所示:

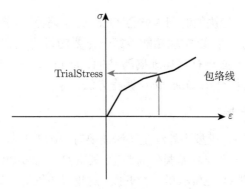

图 2.3.2　应变求应力示意图

2.3.2　代码修改过程

第一步　调试好开发环境.

在进行添加材料以前, 必须确保:

➢ 安装好 visual studio 开发平台
➢ 安装好 Tcl/tk
➢ 下载可靠的 OpenSees 源代码
➢ 在 visual studio 平台正确编译 OpenSees

以上详见第一部分 1.1 下载与运行和第二部分 2.1 下载与编译.

第二步　在 OpenSees 源代码子文件 uniaxial 中找出添加材料模板 New UniaxialMaterial.h 和 NewUniaxialMaterial.cpp, 并在此文件夹中复制并粘贴文件, 同时重命名为 GeneralElastic.h 和 GeneralElastic.cpp. 如图 2.3.3 和图 2.3.4 所示.

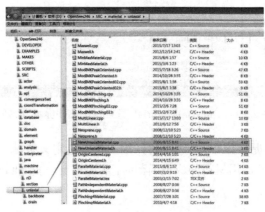

图 2.3.3　找到 NewUniaxialMaterial.h 和 NewUniaxialMaterial.cpp 文件

2.3 OpenSees 添加新材料

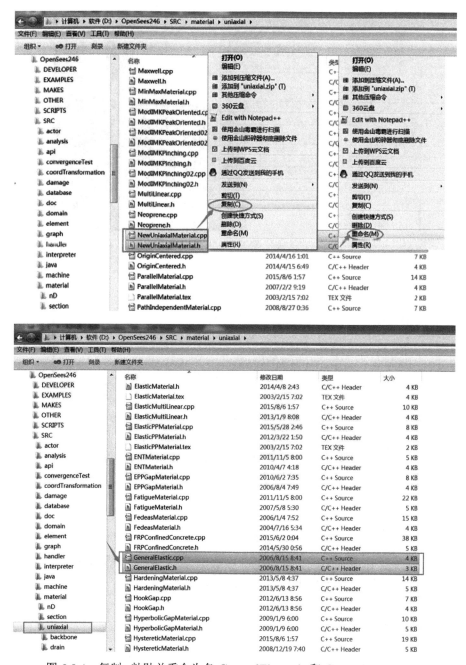

图 2.3.4 复制、粘贴并重命为名 GeneralElastic.h 和 GeneralElastic.cpp

第三步 在 visual studio 中的 OpenSees 文件夹中添加 GeneralElastic.h 和 GeneralElastic.cpp 文件并进行重新编译,确定添加过程正确。如图 2.3.5 所示。

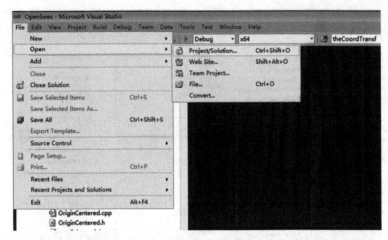
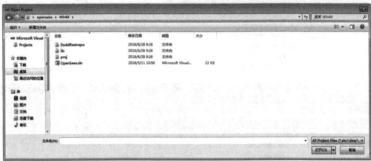
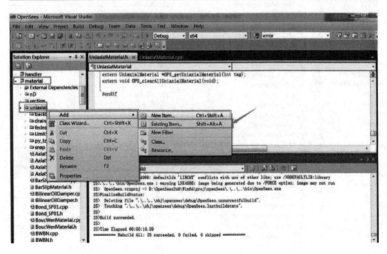

2.3 OpenSees 添加新材料 · 159 ·

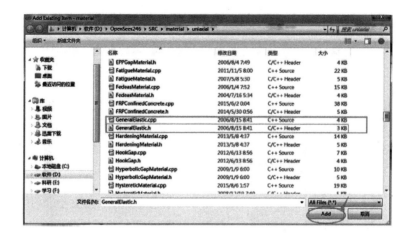

图 2.3.5 uniaxial 添加 GeneralElastic.h 和 GeneralElastic.cpp

第四步 打开 GeneralElastic.h 和 GeneralElastic.cpp 文件，将文件中的 NewUniaxialMaterial 替换为 GeneralElastic，修改 MAT_TAG_GeneralElastic 1976 为 MAT_TAG_GeneralElastic 20160428，并重新编译，确定修改过程语法正确. 如图 2.3.6～图 2.3.8 所示.

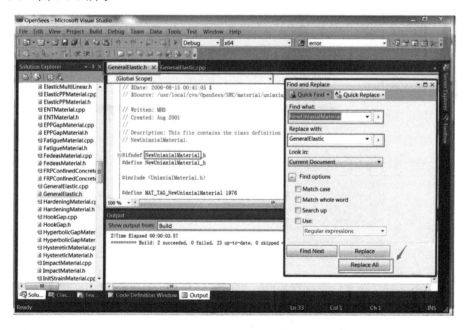

图 2.3.6 将 GeneralElastic.h 文件中 NewUniaxialMaterial 替换为 GeneralElastic

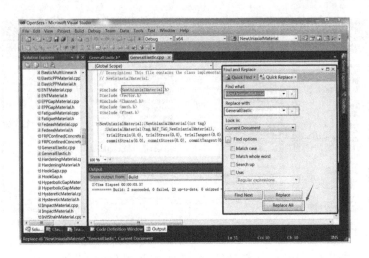

图 2.3.7 将 GeneralElastic.cpp 文件中 NewUniaxialMaterial 替换为 GeneralElastic

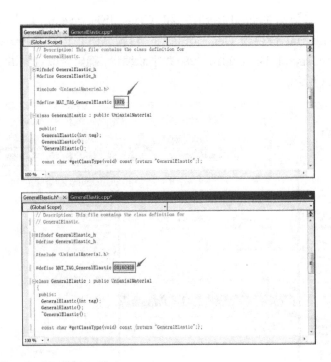

图 2.3.8 修改 GeneralElastic.h 文件中 MAT_TAG_GeneralElastic 数字

第五步 确定 GeneralElastic 材料类的私有数据成员，即修改 GeneralElastic.h 文件中的私有数据成员，如图 2.3.9 所示。

2.3 OpenSees 添加新材料   · 161 ·

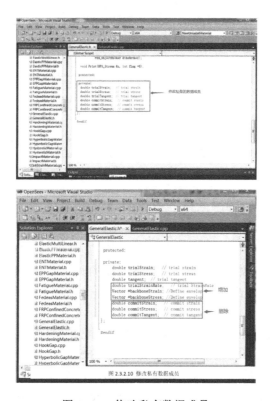

图 2.3.9 修改私有数据成员

第六步 修改 GeneralElastic 类的.h 文件的成员函数声明和实现部分成员函数. 如图 2.3.10 和图 2.3.11 所示.

图 2.3.10 GeneralElastic.h 文件成员函数声明修改前

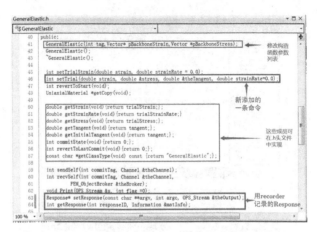

图 2.3.11　GeneralElastic.h 文件成员函数声明修改后

第七步　实现.h 文件中已经声明但未实现的其他成员函数，并重新编译，确定修改过程语法正确. 各个成员函数依次修改后的代码如下所示：

◆ 构造函数

构造函数是当 GeneralElastic 对象生成 (分配内存) 时自动执行的代码，此例中拷贝数组的值给 backboneStress 和 backboneStrain 指向的数组，并调用 revertToStart() 回滚到最初状态 (比如：应力、应变设为 0)。

```
GeneralElastic::GeneralElastic(int tag,Vector* pBackboneStrain,Vector
*pBackboneStress)
    :UniaxialMaterial(tag,MAT_TAG_GeneralElastic),
    trialStrain(0.0), trialStress(0.0),tangent(0.0)
{
if (pBackboneStrain !=0){
    backboneStress = new Vector(*pBackboneStress);
    backboneStrain = new Vector(*pBackboneStrain);
}
else {
    opserr<<"Fatal:  no backbone data in GeneralElasticMaterial!"
<<endln;
    exit(-1);
}
this->revertToStart();
}
```

◆ 析构函数

2.3 OpenSees 添加新材料

由于目前 OpenSees 版权兼容性问题,如果在你的计算机上有错误,可删除这一段代码.

```
GeneralElastic::~GeneralElastic()
{
    if (backboneStrain !=0) { delete [] backboneStrain; backboneStrain = 0; }
    if (backboneStress !=0) { delete [] backboneStress; backboneStress = 0;}
}
```

◆ setTrialStrain()

setTrialStrain() 函数为核心函数之一,它被单元调用. 每个迭代步中,当单元计算此高斯点的应变后,通过调用材料的 setTrialStrain(),计算材料点的应力和刚度,并存在材料的私有成员变量中. 在三维问题中,OpenSees 单元传来的应变为工程应变 (比如; $\gamma_{xy} = 2\varepsilon_{xy}$).

```
int GeneralElastic::setTrialStrain(double strain, double strainRate)
{
double sign = 1.0;
    if (fabs(strain)<1.e-14) sign =0;
    if (strain<-1.e-14) sign=-1;
    trialStrain = strain;
    trialStrainRate = strainRate;
    int i=0;
    while ((i<backboneStrain->Size())&(fabs(trialStrain)>(*backboneStrain)(i)+1.e-14))
       i++;
    if (i==0){trialStress = 0;
       tangent= ((*backboneStress)(1)-(*backboneStress)(0))/((*backboneStrain)(1)- (*backboneStrain)(0)); //interpolation
}
    elseif (i == backboneStrain->Size()) { // too big strain, keep horizontal
       trialStress = sign * (*backboneStress)(i-1);
       tangent = 0;
}
```

```
else { // general case
    trialStress = (*backboneStress)(i-1)+(fabs(trialStrain)-(*backboneStrain)(i-1))
    *((*backboneStress)(i)-(*backboneStress)(i-1))/((*backboneStrain)(i) -(*backboneStrain)(i-1));
    trialStress *= sign;
tangent = ((*backboneStress)(i)-(*backboneStress)(i-1))/((*backboneStrain) (i)- (*backboneStrain)(i-1));
    } if (fabs(tangent)< 1.e-14) tangent =((*backboneStress)(1)-(*backboneStress)(0))/((*backboneStrain)(1)-(*backboneStrain)(0))*1.0e-9;
    if (tangent < -1.e-14) tangent = -((*backboneStress)(1)-(*backboneStress)(0))/((*backboneStrain)(1)-(*backboneStrain)(0))*1.0e-9;
return 0;
}
```

- setTrial()

```
int GeneralElastic::setTrial(double strain, double&stress, double&theTangent, double strainRate)
{
this->setTrialStrain(strain, strainRate);
theTangent = tangent;
stress = trialStress;
return 0;
}
```

- revertToStart()

revertToStart() 将材料状态设为初始值 (往往为零应力、零应变等), 往往被构造函数调用.

```
int GeneralElastic::revertToStart(void)
{
    trialStrain = 0.;
    trialStrainRate = 0.;
    trialStress = 0.0;
```

2.3 OpenSees 添加新材料

```
    tangent = 0.0;
return 0;
}
```

- getCopy()

  getCopy() 将此材料对象本身 (GeneralElastic) 复制一份, 返回新对象的指针. 此函数被单元调用, 单元为每个高斯点复制一个材料对象 (而不是用 New GeneralElastic() 生成). 每个高斯点是一个对象, 独立管理自己的应力、应变等数据.

```
UniaxialMaterial *
GeneralElastic::getCopy(void)
{
    GeneralElastic *theCopy = new GeneralElastic(this->getTag(),backboneStrain,backboneStress);
    theCopy->trialStrain = trialStrain;
    theCopy->trialStrainRate = trialStrainRate;
return theCopy;
}
```

- Print()

```
void
GeneralElastic::Print(OPS_Stream &s, int flag)
{
    s <<"Elastic tag:   "<<this->getTag() << endln;
    s <<" backboneStrain :   "<< *backboneStrain << endln;
    s <<" backboneStress :   "<< *backboneStress << endln;
return;
}
```

- setResponse()

  setResponse() 为 TCL 的 recorder 命令提供解析服务, 规定哪些关键词合法, 并为每个关键词指定一个整数标识 (比如应力 "stress" 赋值 1).

```
Response*
GeneralElastic::setResponse(constchar **argv, int argc, OPS_Stream &theOutput)
{
    Response *res = UniaxialMaterial::setResponse(argv, argc, theOutput);
    if (res != 0) return res;
```

```
else return 0;
}
```

♦ getResponse()

getResponse() 也为 TCL 的 recorder 命令提供解析服务，根据 setResponse() 指定的整数标识返回所需记录的内存变量 (比如读到 1 就返回应力). setResponse() 在每一步 commit 后调用，记录的是每一时步迭代收敛后的值 (节点位移、应力等).

```
int
GeneralElastic::getResponse(int responseID, Information &matInfo)
{
return UniaxialMaterial::getResponse(responseID,matInfo);
}
```

♦ commitState()

commitState() 在本时步收敛后被调用 (隐式解法)，在塑性本构模型中非常重要，把迭代步的 trial 的应力、应变等值赋给 committed 的应力、应变等值，然后准备进入下一个计算时步. 弹性材料不需要，在后面塑性材料介绍中需要. 因此直接返回 0.

第八步 修改 TclModelBuilderUniaxialMaterialCommand.cpp 文件，并重新编译，确定修改过程语法正确. 如图 2.3.12 和图 2.3.13 所示.

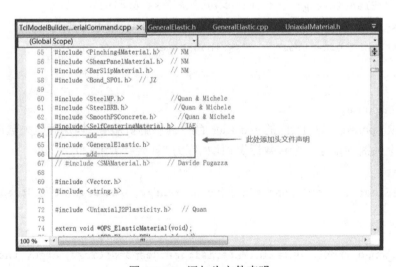

图 2.3.12 添加头文件声明

2.3 OpenSees 添加新材料

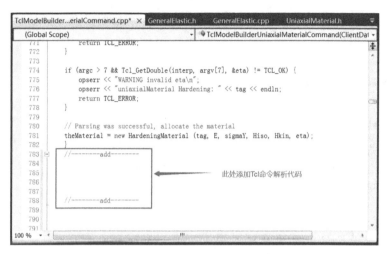

图 2.3.13 添加 Tcl 解析代码

♦ TclModelBuilderUniaxialMaterialCommand.cpp 文件添加如下代码:
```
elseif (strcmp(argv[1],"GeneralElasticMaterial") == 0) {
    if (argc < 3) {opserr <<"WARNING insufficient arguments\n";
    printCommand(argc,argv);
    opserr <<"Want:  uniaxialMaterial GeneralElasticMaterial tag?
-strain {} -stress {}"<< endln;
    return TCL_ERROR;
    }
int tag;
Vector * backboneStrain=0;
Vector * backboneStress=0;
if (Tcl_GetInt(interp, argv[2], &tag) != TCL_OK) {
    opserr <<"WARNING invalid tag in GeneralElastic Material"<< endln;
        return TCL_ERROR;
}
if (strcmp(argv[3],"-strain") == 0) {
    int pathSize;
    TCL_Char **pathStrings;
if (Tcl_SplitList(interp, argv[4], &pathSize, &pathStrings) !=TCL_OK) {
        opserr <<"WARNING problem splitting path list "<< argv[4] <<" -
";
```

```
            opserr <<" stress -values {path} ...  \n";
            return 0;
        }
backboneStrain = new Vector(pathSize);
for (int i=0; i<pathSize; i++) {
double value;
if ( Tcl_GetDouble(interp, pathStrings[i], &value) != TCL_OK) {
   opserr <<"WARNING problem reading path data value "<<
pathStrings[i] <<" - ";
opserr <<" -strain {path} ...  \n";
return 0;
}
(*backboneStrain)(i) = value;
}
}
else {opserr<<"error command in generalElasticMaterial!"<<endln; exit
(-1);}
if (strcmp(argv[5],"-stress") == 0) {
int pathSize;
    TCL_Char **pathStrings;
if (Tcl_SplitList(interp, argv[6], &pathSize, &pathStrings) != TCL_OK)
{
opserr <<"WARNING problem splitting path list "<< argv[4] <<" - ";
opserr <<" stress -values {path} ...  \n";
return 0;
}
backboneStress = new Vector(pathSize);
for (int i=0; i<pathSize; i++) {
double value;
if ( Tcl_GetDouble(interp, pathStrings[i], &value) != TCL_OK) {
opserr <<"WARNING problem reading path data value "<< pathStrings[i]
<<" - ";
opserr <<" -strain {path} ...  \n";
return 0;
}(
```

2.3 OpenSees 添加新材料

```
*backboneStress)(i) = value;
}
}
else {opserr<<"error command in generalElasticMaterial!"<<endln; exit
(-1);}
theMaterial = new GeneralElastic(tag, backboneStrain, backboneStress);
}
```

2.3.3 建立 Tcl 模型，调试程序

第一步 建立 Tcl 模型，并将 Tcl 文件放入 OpenSees 指定的 OpenSees 文件夹中，设置断点，进行 Debug. 如图 2.3.14 所示.

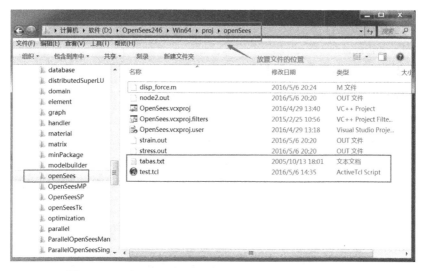

图 2.3.14 将 Test.tcl 和 tabas.txt 文件放在指定文件夹

◆ Tcl 模型代码:

```
wipe ;
model basic -ndm 2 -ndf 2
node 1 0.0 0.0
node 2 10.0 0.0 -mass 10000.0 10000.0
fix 1 1 1
fix 2 0 1
set strainList {0 0.01 0.02 0.03 0.04 0.05 0.07 0.09}
set stressList {0 2 3 3.5 3 2.5 2.0 1.0}
```

uniaxialMaterial GeneralElasticMaterial 1 -strain $strainList -stress $stressList //定义新添加的材料
element truss trussID node1 node2 A matID
element truss 1 1 2 1 1
recorder Node -file node2.out -time -node 2 -dof 1 2 disp
recorder Element -file stress.out -time -ele 1 -material stress
recorder Element -file strain.out -time -ele 1 -material strain
set tabas "Path -filePath tabas.txt -dt 0.02 -factor 140"
pattern UniformExcitation 1 1 -accel $tabas
constraints Transformation
numberer RCM
test NormDispIncr 1.E-8 25 2
algorithm Newton
system BandSPD
integrator Newmark 0.55 0.275625
analysis Transient
analyze 1000 0.01

注:Tcl 建模命令详细说明参见 OpenSees 官网。

第二步 在文件中用鼠标左键或者 F9 设置断点，F5 或者 F10 进行 Debug。如图 2.3.15 和图 2.3.16 所示。

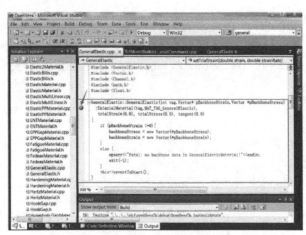

图 2.3.15 设置断点

2.3 OpenSees 添加新材料

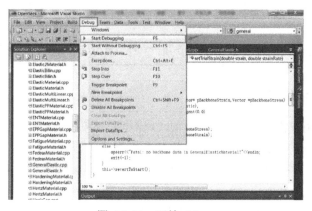

图 2.3.16　开始 debug

第三步　程序调试无误后, 绘制数据图像. 如图 2.3.17 所示.

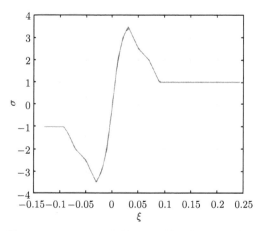

图 2.3.17　Matlab 绘制 truss 单元应力 - 应变图

注意　添加新单元和新材料相对复杂一些, 其难度和在商业有限元 (比如 Abaqus) 中加新单元和新材料是相似的. 由于 OpenSees 开源, 调试更加简单和灵活. 当用户需要实现非常特殊的功能 (比如倒塌模拟中轴向承载力和侧向变形有某种关系) 时, OpenSees 往往能够实现其他有限元无法做到的功能. 在实现此类功能时, 用户第一步是阅读 OpenSees 已有类似的代码, 像学习教科书一样, 充分利用这个资源库, 仿照已有程序 "照猫画虎" 地编写自己的代码, 从而节省时间, 高效地完成任务.

2.4 OpenSees 添加一维理想弹塑性材料

2.4.1 添加新材料背景资料介绍

本节将向用户说明如何在 Windows10 操作系统和 VisualStudio 2015 集成开发环境下编译最新的 OpenSees 源码并添加一维理想弹塑性材料 (以下称为 PerfectPlasticMaterial 类). 不同版本的 Windows 和 C++ 添加材料方法类似.

如同 2.3.1 小节所述, 该 PerfectPlasticMaterial 类也以单轴材料类 (UniaxialMaterial) 为基类, 通过构造几个关键的函数成员 (如 PerfectPlasticMaterial(), setTrialStrain(), commitState() 等), 实现新材料的功能 (图 2.4.1).

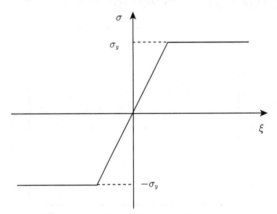

图 2.4.1 1D PerfectPlasticMaterial 本构关系

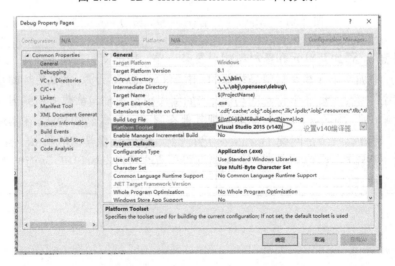

2.4 OpenSees 添加一维理想弹塑性材料

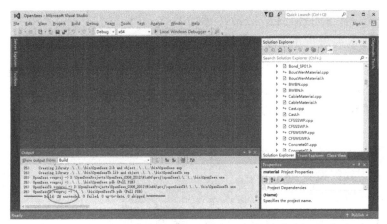

图 2.4.2　OpenSees 最新源码在 VS2015 平台编译成功

2.4.2　配置开发环境

1) 将用户的操作系统升级至 Windows 10；
2) 下载最新 OpenSees 源码 (参照 2.1.1 小节)；
3) 下载并安装对应版本 TCL(参照 2.1.2 小节)；
4) 下载并安装 VisualStudio 2015 集成开发环境 (参照 2.1.3 小节、2.1.4 小节)；
5) 在 VisualStudio 2015 平台正确编译最新 OpenSees 源码 (参照 2.1.5 小节)。

2.4.3　代码修改过程

第一步　在源代码子文件 uniaxial 中找到基类文件 NewUniaxialMaterial.h 和 NewUniaxialMaterial.cpp, 复制该文件并重命名为 PerfectPlasticMaterial.h 和 PerfectPlasticMaterial.cpp (图 2.4.3)。

图 2.4.3　复制文件并重命名为 PerfectPlasticMaterial.h 和 PerfectPlasticMaterial.cpp

第二步 在 Solution Explorer 中添加 PerfectPlasticMaterial.h 和 PerfectPlasticMaterial.cpp 并进行重新编译，确认添加过程正确，如图 2.4.4 所示。

第三步 打开 PerfectPlasticMaterial.h 和 PerfectPlasticMaterial.cpp 文件，将文件中的 NewUniaxialMaterial 替换为 PerfectPlasticMaterial，并修改 1976 为 20170308，重新编译，确定修改过程语法正确，如图 2.4.5 所示。

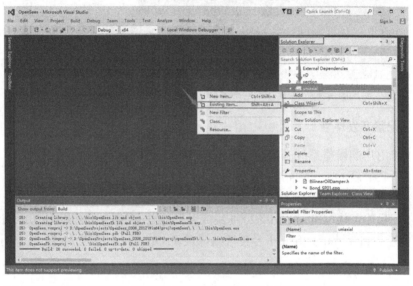

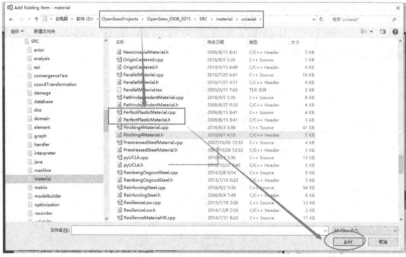

图 2.4.4 将 PerfectPlasticMaterial 类添加至 material 工程

2.4 OpenSees 添加一维理想弹塑性材料

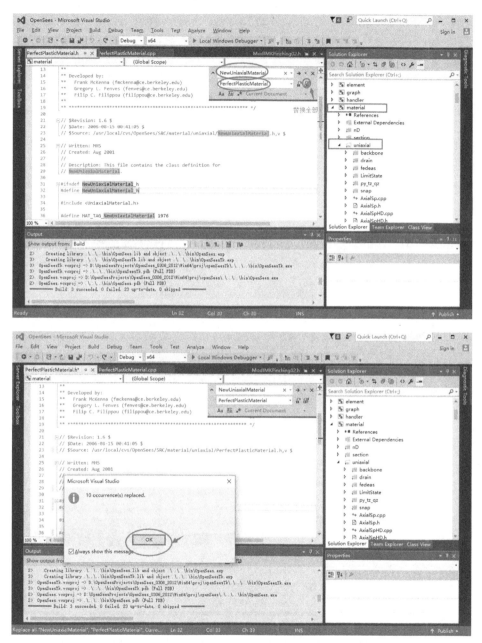

图 2.4.5 将 PerfectPlasticMaterial.h 文件中 NewUniaxialMaterial 替换为 PerfectPlasticMaterial

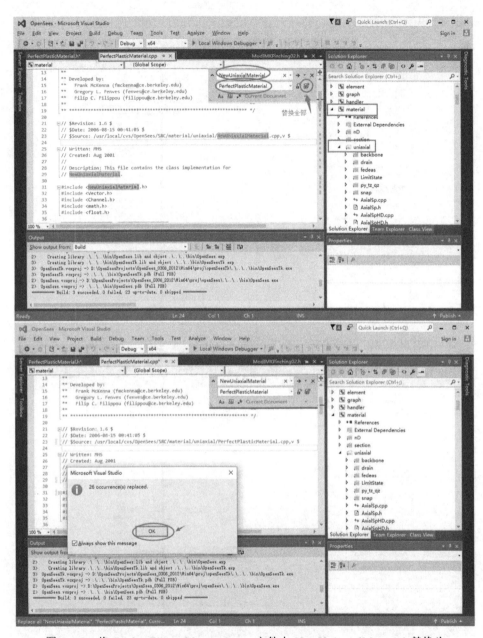

图 2.4.6 将 PerfectPlasticMaterial.cpp 文件中 NewUniaxialMaterial 替换为 PerfectPlasticMaterial

2.4 OpenSees 添加一维理想弹塑性材料

图 2.4.7　修改 PerfectPlasticMaterial.h 文件中 MAT_TAG_PerfectPlasticMaterial 数字

第四步　确定 PerfectPlasticMaterial 材料类的私有数据成员，即修改 Perfect-PlasticMaterial.h 文件中的私有数据成员，如图 2.4.8 所示。

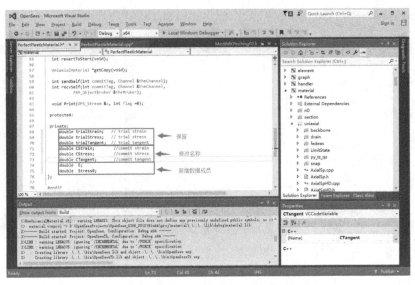

图 2.4.8 修改和新增私有数据成员

第五步 确定 PerfectPlasticMaterial 材料类的函数成员，即修改和添加.h 文件的成员函数声明，如图 2.4.9 所示.

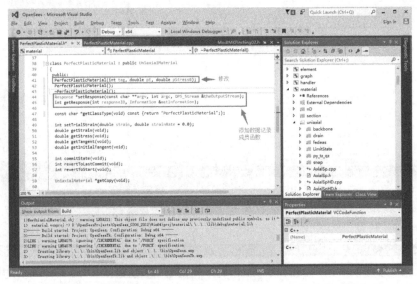

图 2.4.9 修改和新增 PerfectPlasticMaterial.h 文件成员函数声明

第六步 实现.h 文件中已经声明的成员函数，并重新编译，确定修改过程语法正确，各个成员函数依次修改或添加的代码如下所示.

◆ `PerfectPlasticMaterial()`

2.4 OpenSees 添加一维理想弹塑性材料

```cpp
PerfectPlasticMaterial::PerfectPlasticMaterial(int tag,double
pE,double pStress0)
  :UniaxialMaterial(tag,MAT_TAG_PerfectPlasticMaterial),
   trialStrain(0.0), trialStress(0.0), trialTangent(0.0)
{
  E = pE;
  Stress0 = pStress0;
  this->revertToStart();
}
```

◆ setTrialStrain()

```cpp
int
PerfectPlasticMaterial::setTrialStrain(double strain, double strainRate)
{
        // elastic predictor
               trialStrain = strain;
               double dStrain = trialStrain - CStrain;
               trialStress = CStress + E*dStrain;
               trialTangent = E;

        // plastic corrector
               double eps = 1e-14;
               if(trialStress > Stress0)
                   {trialStress =  Stress0;trialTangent = eps;}
               if(trialStress < -Stress0)
                   {trialStress =  -Stress0;trialTangent = eps;}
  return 0;
}
```

◆ commitState ()

在每个时步 (通常包括几个迭代步) 收敛后, 执行此操作. 让存 commit 的变量同步为存 trail 的变量, 然后准备下一时步计算.

```cpp
int
PerfectPlasticMaterial::commitState(void)
{
    //record commit value
    CStrain  = trialStrain;
```

```
    CStress = trialStress;
    CTangent = trialTangent;
    return 0;
}
```
◆ revertToStart()
```
int
GeneralElastic::revertToStart(void)
{
    trialStrain = 0.;
    trialStress = 0.0;
    trialTangent = 0.0;
    CStrain = 0.;
    CStress = 0.0;
    CTangent = 0.0;
    return 0;
}
```
◆ getCopy()
```
UniaxialMaterial *
PerfectPlasticMaterial::getCopy(void)
{
    PerfectPlasticMaterial *theCopy = new PerfectPlasticMaterial(this->
    getTag(),E,Stress0);
    return theCopy;
}
```
◆ setResponse()
```
Response*
PerfectPlasticMaterial::setResponse(const char **argv, int argc,
OPS_Stream &theOutput)
{
    Response *res = UniaxialMaterial::setResponse(argv, argc,
    theOutput);
    if (res != 0)        return res;
    else    return 0;
}
```
◆ getResponse()

2.4 OpenSees 添加一维理想弹塑性材料

```
int
PerfectPlasticMaterial::getResponse(int responseID, Information
&matInfo)
{
  return UniaxialMaterial::getResponse(responseID,matInfo);
}
```

◆ setParameter() `PerfectPlasticMaterial::setParameter(const char **argv, int argc, Parameter &param) {`
```
    if (strcmp(argv[0], "E") == 0) {
        param.setValue(E);   // size of parameter
        return param.addObject(1, this);
    }
    return -1;
}
```

◆ updateParameter() `int PerfectPlasticMaterial::updateParameter(int parameterID, Information &info) {`
```
    switch (parameterID) {
    case 1:
        this->E = info.theDouble;
        trialTangent = E;
        break;
    default:
        return -1;
    }
    return 0;
}
```

> 更新材料参数的功能可通过 setParameter() 和 updateParameter() 接口实现，对应 Tcl 命令应在分析命令 analyze 之间写入，格式如下：
>
> parameter 1
> addToParameter 1 element 1 material E

updateParameter 1 40.0

注：addToParameter 命令中 "element 1 material E" 的写法可以参考 recorder element(参考第 40 页) 逐级解析原则，即从单元 -> 截面 -> 材料的逐级调用 setResponse() 函数。

第七步 修改 TclModelBuilderUniaxialMaterialCommand.cpp 文件，并重新编译，确定修改过程语法正确，如图 2.4.10–图 2.4.12 所示。

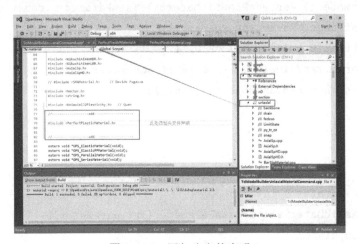

图 2.4.10 添加头文件声明

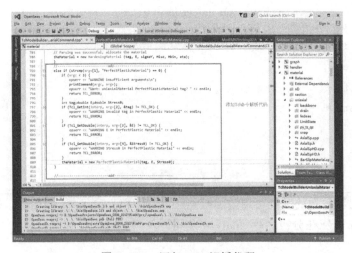

图 2.4.11 添加 Tcl 解析代码

2.4 OpenSees 添加一维理想弹塑性材料

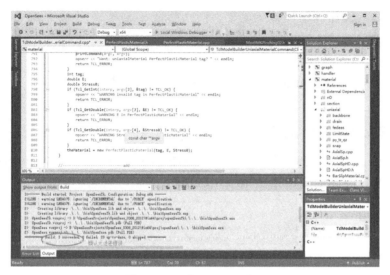

图 2.4.12　确认无语法错误

◆ TclModelBuilderUniaxialMaterialCommand.cpp文件添加如下代码：
```
else if (strcmp(argv[1],"PerfectPlasticMaterial") == 0) {
if (argc < 3) {opserr << "WARNING insufficient arguments\n";
    printCommand(argc,argv);
    opserr << "Want: uniaxialMaterial PerfectPlasticMaterial
        tag? " << endln;
    return TCL_ERROR;
}
int tag;double E;double Stress0;
if (Tcl_GetInt(interp, argv[2], &tag) != TCL_OK) {
    opserr << "WARNING invalid tag in PerfectPlasticity"
        << endln;
    return TCL_ERROR;
}
if (Tcl_GetDouble(interp, argv[3], &E) != TCL_OK) {
    opserr << "WARNING E in PerfectPlasticity Material" << endln;
    return TCL_ERROR;
}
if (Tcl_GetDouble(interp, argv[4], &Stress0) != TCL_OK) {
    opserr << "WARNING Stress0 in PerfectPlasticity"
```

```
        << endln;
        return TCL_ERROR;
}
theMaterial = new PerfectPlasticMaterial(tag,E,Stress0);
}
```

2.4.4 建立 Tcl 模型,调试程序

第一步 建立 Tcl 模型,并将 Tcl 文件放入 OpenSees 指定的 openSees 文件夹中,设置断点,进行 Debug,如图 2.4.13 所示.

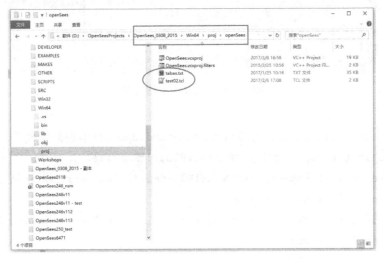

图 2.4.13 将 Test02.tcl 和 tabas.txt 文件放在指定文件夹

◆ Tcl 模型代码:

```
wipe ;
model basic -ndm 2 -ndf 2
node 1   0.0   0.0
node 2   10.0  0.0    -mass 10000.0  10000.0
fix 1 1 1
fix 2 0 1
#uniaxialMatyerial PerfectPlasticMaterial $tag      $E      $Stress0
uniaxialMaterial PerfectPlasticMaterial     1    2e1     3.0e-2
uniaxialMaterial Elastic                    2    20.0
element truss      1     1    2    1     1
element truss      2     1    2    1     2
```

2.4 OpenSees 添加一维理想弹塑性材料

```
recorder Node -file node2.out -time -node 2 -dof 1 2 disp
recorder Element -file stress1.out  -time  -ele 1  -material stress
recorder Element -file strain1.out  -time  -ele 1  -material strain
recorder Element -file stress2.out  -time  -ele 2  -material stress
recorder Element -file strain2.out  -time  -ele 2  -material strain
set tabas "Path -filePath tabas.txt -dt 0.02 -factor 4"
pattern UniformExcitation 1   1 -accel       $tabas

constraints Transformation
numberer RCM
test NormDispIncr 1.E-8 25  2
algorithm Newton
system BandSPD
integrator Newmark 0.55 0.275625
# (0.55+0.5)^2/4=0.275625
analysis Transient
analyze 1000  0.01
```

注:Tcl 建模命令详细说明参见 OpenSees 官网.

第二步　在文件中用鼠标左键或者 F9 设置断点, F5 或者 F10 进行 Debug. 如图 2.4.14 和图 2.4.15 所示.

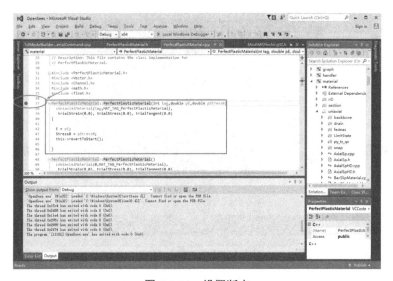

图 2.4.14　设置断点

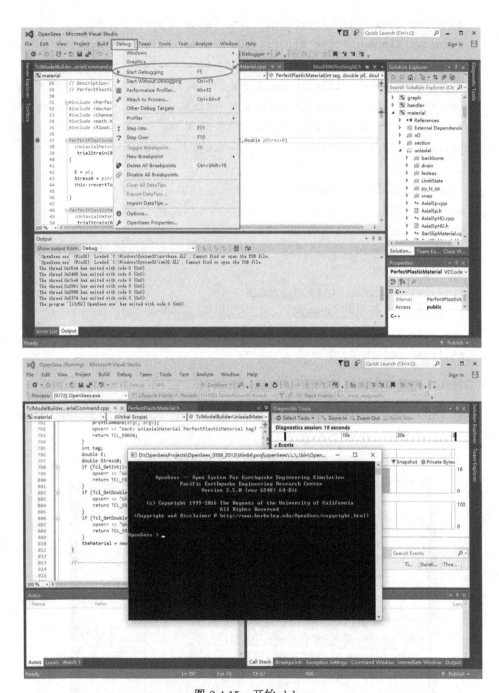

图 2.4.15 开始 debug

第三步 程序调试无误后,绘制数据图像. 如图 2.4.16 所示.

2.4 OpenSees 添加一维理想弹塑性材料

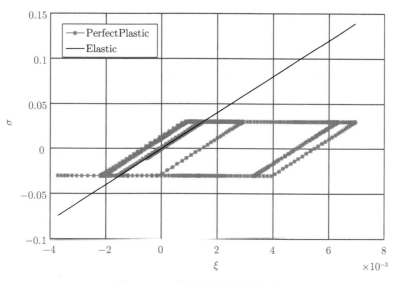

图 2.4.16 绘制材料点滞回曲线

索 引

下载与运行　1
安装路径　2
建模　4
　　wipe　5
　　node　5
　　element　5
　　fix　5
　　节点与单元质量单元与材料密度　25,69
加载
　　load control 与 pattern　7,52,54
　　位移控制 displacement control　27
　　load Const –time 0　8,53
　　静力、动力分析　5
　　重力分级施加　24
　　基底一致激励　26,27
　　多点约束激励　27,97
地震输入文件　8,9
　　rayleigh 阻尼、数值阻尼　26,53
　　eigen 分析　25,26
Tcl 语法
recorder 记录单元与材料信息　40
recorder 记录支座与节点反力 reaction　8,97
梁柱框架结构　21
　　二维、三维局部坐标系定义　30,38,49
　　section Aggregator 使用　30
　　力插值梁柱单元　31

坐标转换与几何大变形　36
　　纤维截面 fiber patch layer　38,46
　　concrete01 与 steel01 材料　37
土–结相互作用　55
　　地基土类型　61
　　桩、土结点连接与土边界　70,72
流固耦合　72
　　人工透射边界　72
　　流体单元、流体边界　73
　　截断 Drucker-prager 模型　74,77
饱和砂土液化　79
　　Bounding surface 模型　81
　　multi-yield surface 模型　81
　　updateMaterialStage　81
数值优化　85
　　设计变量约束函数 OpenSees-SNOPT 优化　85,89,92
OpenSees 与其他软件集成
　　CS 技术　93,95
OpenSees 前后处理 GID　100
OpenSees 编程　120
　　添加非线性弹性材料　120
　　添加理想弹塑性材料　172
C++ 简介　131
parameter、updateParameter　181